The Event Horizon: Homo Prometheus and the Climate Catastrophe

In memory of Colin Peter Groves
24 June 1942–30 November 2017

Andrew Y. Glikson

The Event Horizon: Homo Prometheus and the Climate Catastrophe

Springer

Andrew Y. Glikson
Research School of Earth Science
Australian National University
Canberra, ACT, Australia

ISBN 978-3-030-54736-3   ISBN 978-3-030-54734-9   (eBook)
https://doi.org/10.1007/978-3-030-54734-9

Cover illustration: Prometheus attacked by the eagle and Sisyphus carrying a boulder. A Greek vase. https://en.wikipedia.org/wiki/Theogony#/media/File:Atlas_Typhoeus_Prometheus.png

This Springer imprint is published by the registered company Springer Nature Switzerland AG
The registered company address is: Gewerbestrasse 11, 6330 Cham, Switzerland

Foreword

Human history and Earth history are colliding at an accelerating rate. We have now entered the 'end game'. Do the cataclysms of Earth's past tell us what future our children and grandchildren might face? The answer is imminent. Professor (William Steffen, Emeritus Professor in climate science, Fenner School of Environment & Society, Australian National University). *Andrew Glikson is a superb writer, philosopher and scientist. There is an enormous gulf between The Event Horizon: Homo Prometheus and the climate catastrophe and every other book about the consequences of climate change that exists, that gulf being the level of research that has gone into it. Glikson has an extraordinary grasp of the scientific literature and uses it to full effect. This book may not be for everybody as it is moderately technical, but it is for anybody with a science background and anybody else who can accommodate scientific information. For these people, this book is an easy read and is thoroughly engaging. This is a gem of a book—a must read for anybody who wants to know about what humanity is doing to the global environment and what the consequences are likely to be.* (Prof. John Veron, former chief scientist, the Australian Institute of Marine Science). *This is a very timely book written by prominent and prolific global scholar, Dr. Andrew Glikson. In this book, Andrew takes the reader on a challenging journey from human scales and perspectives on that of our collapsing planet. By synthesizing past events (deep time) to that of present-day changes to our world from humans, Andrew weaves a captivating perspective of what lies ahead. The ramifications of the next phase of human history are truly shocking—keeping the reader on the edge of her/his seat. It is deeply evidence based, well written and compelling—and includes events up to the moment, such as the implications of escalation of Australia's fires over the past three months* (Prof. Ove Hoegh-Guldberg, Professor of Marine Science, the University of Queensland). *The list of chapters displays the breadth and depth of the author's understanding of Earth's geological environmental crises through to the rise of Homo sapiens and our apparent inability to control our (reptilian) brain with its irrational emotions and lack of consequential realization. This book leads us inevitably to the event horizon and an unpredictable future. Can we wake up in time to reduce the extreme consequences of our failings?* (Prof. Victor Gostin, Hon A/Professor, the University of Adelaide).

Canberra, ACT, Australia   Dr. Andrew Y. Glikson

Prologue

This book provides an account of the moment of truth Homo sapiens is facing, as the Bulletin of the Atomic Scientists has set the doomsday clock to 100 seconds to mid-night, the closest call since 1947, in view of the looming nuclear and climate change threats. Technologically supreme, the species manifests its dreams and its nightmares through science, art, adventures and murderous wars, a paradox symbolized by a candle lighting the dark yet burning itself to extinction, as discussed in the book. The human predicament is signified by the Titan Prometheus, who gave fire and thereby power to the species, paying the ultimate price in eternal suffering. As contrasted with Stapledon's (1972) *Last and First Man*, portraying an advanced species mourns the fate of Earth, *Homo sapiens* proceeds to transfer every extractable molecule of carbon from the Earth into the atmosphere, the lungs of the biosphere, committing a demise of the planetary life support system. It further transfers derived plastic and micro-plastic into the hydrosphere, severely harming marine life and birds. As these lines are written fires are burning in several continents, the Earth's ice sheets are melting and the oceans are rising, threatening to inundate the planet's coastal zones and river valleys, where civilization has arisen and humans live and grow food. With the exception of birds such as Hawks, Black Kites and Fire Raptors, humans constitute the only life form utilizing fire, developing processes they can hardly control. For longer than one million years, gathered around camp fires during the long nights, mesmerized by the flickering life-like dance of the flames, prehistoric humans acquired imagination, yearnings for omnipotence, premonitions of death, cravings for immortality and concepts of the super-natural. Humans live in realms of perceptions, dreams, myths and legends, in denial of critical facts, waking up for a brief moment to witness a world as beautiful as it is cruel. Existentialist philosophy allows a way of coping with the unthinkable. Looking into the future generates fear, an instinctive sense arising in animals once endangered, but which is obsessing the human mind, generating a conflict between the intuitive reptilian brain and a growing neocortex.

Acknowledgements I am grateful to Brenda McAvoy for proof reading of the manuscript, Will Steffen for scientific discussions, Ove Guldberg, Victor Gostin, John Veron, Helen Caldicott, Robert Manne and Hugh Davies for reviews and comments.

Contents

About the Author

Andrew Y. Glikson An Earth and paleo-climate scientist, studied geology at the University of Jerusalem and graduated at the University of Western Australia in 1968. He conducted geological surveys of the oldest geological formations in Australia, South Africa, India and Canada; studied large asteroid impacts, including their effects on the atmosphere and oceans and the mass extinction of species. Since 2005 he studied the relations between climate and human evolution. He was active in communicating nuclear issues and climate change evidence to the public and to parliament through papers, lectures, conferences and presentations.

1

Introduction: From Homo Prometheus to Terra Incognita

During times of universal deceit, telling the truth becomes a revolutionary act.
George Orwell

Sunset's chariot drowns twilight
The Southern Cross kindles bright
While the moon is still asleep
An aurora lights the deep
Far beyond my sheltered campfire
Nests among the brown hills attire
Where tonight I'm stranded, aching
Within a cage of my own making
Questions burn in me:
Tell me, for whatever reason
Have I emerged into this season
Where will I be once the embers burn
When Earth completes another turn
About its axis, will life exist?
Will desert's bloom the dark resist
Once Gaea makes yet one more flight
Around the star which sets tonight?

This book is concerned with the threats to the terrestrial environment and life emanating from both, the climate catastrophe, where the train has already left the station, and the rising prospects of nuclear war, as expressed in the 2020 statement by the Atomic Scientists, which reads:

Fig. 1.1 Greta Thunberg—A voice on behalf of future generations. *Creative Commons Attribution-Share Alike 4.0 International license*

Humanity continues to face two simultaneous existential dangers—nuclear war and climate change—that are compounded by a threat multiplier, cyber-enabled information warfare, that undercuts society's ability to respond. The international security situation is dire, not just because these threats exist, but because world leaders have allowed the international political infrastructure for managing them to erode.

Lately a voice has emerged, of Greta Thunberg (Fig. 1.1), a teenage girl representing the young of future generations bound to suffer from the criminal blindness that has taken over the world.

Although both the climate crisis and the nuclear peril endanger life on Earth, they arise from somewhat distinct factors: The first from a collective blindness to the consequences of changing the composition of the atmosphere. The second representing the culmination of murderous tribal wars rooted deep in human history. As evidenced by the lack of concern by extreme right and left ideologies, both are linked to unconscious life-negating forces.

An inverse relation may exist between a species' level of consciousness and its longevity once it learns to control fire and creates processes it cannot control. If looking directly into the sun can result in blindness then, according

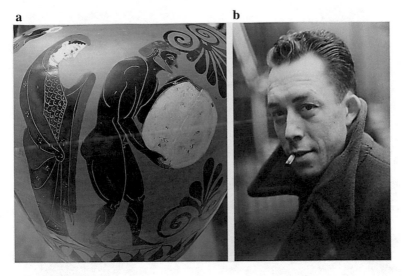

Fig. 1.2   a Sisyphus (Persephone supervising Sisyphus in the underworld, Attica black-figure amphora, c. 530 BC; b Albert Camus (Creative Commons) (Gopnik 2012). Abjuring abstraction and extremism, Camus found a way to write about politics that was sober, lofty, and a little sad. Photograph by Henri Cartier-Bresson/Magnum

to as yet little-understood principles, the deep insights that humans have gained into nature may bear a terrible price. Existentialist philosophy allows a perspective on, and a way of coping with, the unthinkable. Individual ethical and cultural attributes rarely govern the behavior of societies, let alone nations or an entire species. And although the planet may not shed a tear for the demise of civilization, hope on an individual scale is still possible for a fleeting moment in the sense of existentialist philosophy. Going through the black night of the soul, members of the species may be rewarded by the emergence of a conscious dignity, devoid of illusions, recognizing the nature of the absurd and grateful for the glimpse at the universe humans are privileged to observe. "*Having pushed a boulder up the mountain all day, turning toward the setting sun, we must consider Sisyphus happy*" (Albert Camus 1942) (Fig. 1.2).

In his novel *Last and First Man* Stapledon (1972) portrays an advanced human species mourning the fate of the Earth as it is heated by a red giant sun growing toward the planet's eventual demise. As individuals plunge into despair, the species tries to spread itself through the Milky Way disseminating spores containing human genes. By sharp contrast *H. sapiens*, since the 20th and through the early 21st century, is transferring every extractable molecule of carbon from the Earth crust to the atmosphere, perpetrating the demise of the planetary life support system (Schellnhuber 2009).

Fig. 1.3 Bushmen, Botswana, starting a fire by rubbing sticks together

As these lines are written fires are burning in several continents (NASA 2019), the Earth's ice sheets are melting and the oceans are rising, threatening to inundate the planet's coastal zones and river valleys, where civilization arose and a majority of humans live and grow food. A criminal cabal of fossil fuel executives, billionaires and their political and media mouthpieces, aided by a tiny group of fraudulent pseudoscientists (Conway and Orskes 2010), is misrepresenting the science ignoring the basic laws of physics and empirical observations. What is it which renders a technologically brilliant species to be so helpless in securing its future and that of nature on which it depends?

With the exception of birds like Hawks, black kites and fire raptors, which carry burning twigs and (Pickrel 2018) start fires (Fig. 1.3) clearing areas for prey, no other organism is known to use fire. The mastery of fire by *Homo* rendered it a unique genus, gaining powers beyond its physical attributes. For longer than one million years, gathered around camp fires during the long nights, mesmerized by the flickering life-like dance of the flames, prehistoric humans acquired imagination, yearnings for omnipotence, premonitions of death, fears, cravings for immortality and the concepts of the supernatural (Glikson and Groves 2016). The ability to look ahead, generating fear, an instinctive sense arising in animals mainly or only once endangered, has obsessed the human mind, driving it to anticipate real and potential threats. This generated a conflict between the intuitive reptilian brain and a growing

Fig. 1.4 Sculpture in the Great Ballcourt at Chichen Itza depicting sacrifice by decapitation. The figure at left holds the severed head of the figure at right, who spouts blood in the form of serpents from his neck

neocortex (Komninos 1998), often allowing the emotional to supersede the need for survival.

Prehistoric Pantheists have revered the Earth, the rocks, plants and living creatures, whereas the ascent of monotheistic religions has focused worship on sky gods and Olympian deities removed from Earth. The rise of civilization associated with more favorable Holocene climate conditions in the Neolithic (10,000–4,500 BC) enhanced the production of food, allowing humans to express both dreams and nightmares through the construction of monuments and burial rites for the after-life, such as the Egyptian pyramids and Chinese imperial burial caves. Fear and aggression harbored in the reptilian brain generated blood cults (Fig. 1.4) aimed at appeasing the gods through the unleashing of wars.

It is relevant to refer to the Aztecs as an example of human sacrifice dedicated to the gods. Sacrificial victims were often selected from captive warriors, with warfare commonly conducted for the purpose of capturing candidates for sacrifice, the so-called 'flowery war' (xochiyaoyotl), where the eastern Tlaxcala province was a favorite hunting-ground. Those who fought the most bravely or were the most handsome were considered the best candidates for sacrifice to please the gods. Indeed, human sacrifice was particularly reserved for those victims most worthy and was considered a high honor, a direct communion with a god. The Aztecs were a culture obsessed with death: they believed that human sacrifice was the highest form of karmic healing

Fig. 1.5 Yuli Borisovich Khariton (1904–1996) a Russian physicist who is widely credited as being one of the leading scientists in the Soviet Union's nuclear bomb program

(Flanagin 2015; Cartwright 2018). When the Great Pyramid of Tenochtitlan was consecrated in 1487 the Aztecs recorded that 84,000 people were slaughtered in four days.

With the invention of combustion and electromagnetic power ancient blood sacrifices have been transformed into generational sacrifices in industrial-scale wars, with victims of industrial wars on the scale of tens of millions, the splitting of the atom that heralds killings in the order of billions (Fig. 1.5). Subconscious premonitions of suicidal conflict, expressed by Robert Oppenheimer watching the Trinity atomic test fireball on July 16, 1945, read: "*Now I am become Death, the destroyer of worlds*" (from the *Bhagavad Gita*).

While humans continue to be captivated by virtual reality displayed by electronic screens, brave new-world nightmares emerge which artists are the first to sense, as in Bob Dylan's poem "*All along the watchtower*".

> *But you and I we've been through that*
> *And this is not our fate*
> *So let us not talk falsely now*
> *The hour is getting late*

2

Greenhouse Gases and Mass Extinctions

"Earth's rate of global warming is 400,000 Hiroshima bombs a day" (James Hansen, NASA's former chief climate scientist, 2012) … "Earth is now substantially out of energy balance. The amount of solar energy that Earth absorbs exceeds the energy radiated back to space. The principal manifestations of this energy imbalance are continued global warming on decadal time scales and continued increase in ocean heat content" (Hansen 2018).

Several mass extinction events in the history of Earth were associated with fast rises in the level of the main greenhouse gases—CO_2, transient H_2O vapor, CH_4, N_2O, Chlorofluorocarbons, low atmospheric ozone. Past extinctions triggered by volcanic and asteroid impact events involved ejecta fallout, toxic aerosols, aerosol-induced darkening and cooling ("asteroid winter", Hoare 2018), extensive fires, air blasts and tsunami. Atmospheric heating associated with long-lived greenhouse gases released from impact or volcanic activity and lasting in the order of up to 10^4 years (Eby 2009; Solomon 2009) has led to high CO_2 levels (Fig. 2.1) rising at high rates (Fig. 2.2). The Cenozoic era—66 Ma to the present commenced with a post K–T impact freeze followed by greenhouse warming effects, culminating with the Paleocene-Eocene hyperthermal (PETM) at ~55.9 Ma. Consequent warming by more than +5 °C over $15–20 \times 10^3$ years ended with long term cooling during the Eocene, and a sharp plunge in CO_2 and temperature toward formation of the Antarctic ice sheet about 32 Ma (Zachos 2001) (Figs. 2.3 and 2.4).

The subsequent era, dominated by the Antarctic ice sheet, included limited thermal rises in the end-Oligocene and mid-Miocene, followed by Pleistocene

A. Y. Glikson, *The Event Horizon: Homo Prometheus and the Climate Catastrophe*, https://doi.org/10.1007/978-3-030-54734-9_2

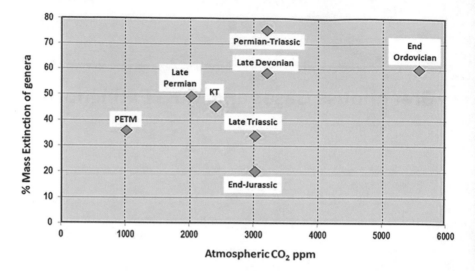

Fig. 2.1 Relations between mass extinction of genera and atmospheric CO_2 levels. Data cited from Royer et al. (2001) and from Keller (2005) and Wignall and Twitchett (2002)

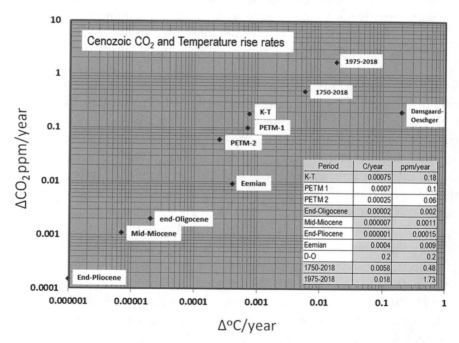

Fig. 2.2 Cenozoic CO_2 and temperature rise rates. Values cited from Table 5.1

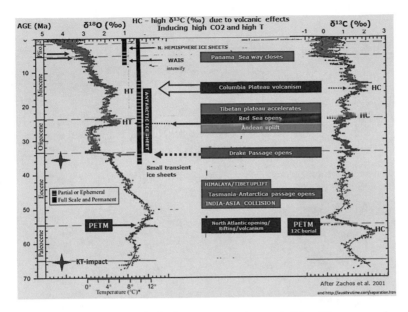

Fig. 2.3 Evolution of the global climate over the past 65 million years (modified after Zachos et al. 2008)

glacial-interglacial cycles and an anthropogenic hyperthermal event mostly post 1750 AD. The greenhouse gas and temperature rise rates of the anthropogenic hyperthermal event exceed those of the PETM and are the highest since the K-T impact event (Table 5.1).

2.1 The K–T Impact-Triggered Hyperthermal Event

The onset of the Cenozoic was marked by the K–T boundary (64.98 ± 0.05 Ma) K–T asteroid impacts (Alvarez et al. 1980), with effects lasting within 32,000 years and atmospheric carbon cycle perturbation less than 5000 years (Renne et al. 2013). These events mark the second largest mass extinction of species in Earth history, when some 46% of living genera disappeared (Keller 2005). The identified parent craters of the K–T event include Chicxulub (~170 km in diameter, Yucatan Peninsula, Mexico) and Boltysh, Ukraine (~25 km in diameter; 65.17 ± 0.64 Ma), Both formed during an active volcanic period which saw continuing eruptions in the Deccan volcanic province (northwest India), dated by U–Pb ages (Schoene et al. 2015) to have commenced approximately ~250,000 years before the K–T boundary age. The volcanism produced >1.1 million km³ of basalt during

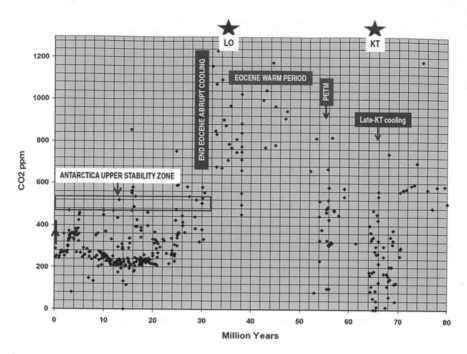

Fig. 2.4 CO_2 levels since 80 Ma. Red arrow indicates the current rise in CO_2. Paleo-CO_2 concentrations based on planktonic foraminifera $d^{13}C$ and stomata indices. Note the upper stability limit of the Antarctic ice sheet of approximately 500 ppm (after Zachos et al. 2001). Stars represent asteroid impact events. LO—Late Eocene impacts; K–T—Cretaceous–Tertiary boundary impacts. Data based on a compilation by D. Royer

~750,000 years, inducing environmental changes preceding the terminal effects of the extraterrestrial bombardment.

The impacts occurred at a time of moderate to low global atmospheric CO_2 concentrations of approximately 600 ppm (Fig. 2.4). Stomata leaf pore-based estimates of atmospheric CO_2 during these events indicate an abrupt rise from ~350 to 500 ppm to at least ~2300 ppm within about 10,000 years, consistent with the transfer of ~4,600 Gigaton Carbon (GtC) to the atmosphere (Beerling et al. 2002). The question of the rise rate of CO_2 is complicated since the gas is released over different time scales, including release from the impacted rocks, fires, warming sea water and other processes, including continuous release of gas from the craters and from ejecta. Climate models suggest consequent forcing of 12 W/m^2, sufficient to warm the Earth's surface by ~7.5 °C in the absence of counter forcing by sulphate aerosols (Beerling et al. 2002). According to these authors CO_2 and temperature rises took place at rates of ~0.18 ppm/year, which is more than an order of magnitude less than the present ~2 to 3 ppm/year.

Short term effects of the K–T asteroid impact include incineration of large land surfaces from the heat pulse of the incoming projectile, from the explosion and settling of hot ejecta air blasts, tsunami and fires (Wolbach et al. 1990), ejection of dust and water vapor, oxidation of atmospheric nitrogen and consequent ozone depletion. Longer term effects included release of CO_2 and other greenhouse gases from impacted carbonates and shales with consequent warming, ocean acidification and anoxia (Covey et al. 1994; Toon et al. 1997).

2.2 The Paleocene-Eocene Thermal Maximum (PETM)

The Paleocene-Eocene Thermal Maximum (PETM) is arguably the best ancient analog of modern climate change. A sharp warming event is represented by the Paleocene-Eocene thermal maximum (PETM) at 56.09 ± 0.03 Ma triggering the largest deep-sea mass extinction event in the last 56 million years as well as a diversification of life in the surface ocean and on land (Pagani et al. 2006; Zachos et al. 2008; Zeebe et al. 2009; Cui et al. 2011; McInerney and Wing 2011).

Zeebe et al. (2009) suggest an increase in CO_2 from pre-PETM levels of 1000 ppm to 1700 ppm during the main phase of the PETM from an initial input of 3000 GtC of methane clathrates of $\delta^{13}C$ lighter than $-50‰$, claiming an increase from 500 ppm to ~850 ppm. This result is similar to Cui et al. (2011) who suggested a contribution of 2500 GtC from methane clathrates of $\delta^{13}C$ of $-60‰$, with an increase in CO_2 from 835 ppm to ~1500 ppm.

McInerney and Wing (2011) suggest carbon release and an input of more than 2000 GtC into the atmosphere, lasting less than 20,000 years, with duration of the whole event of ~200,000 years and a global temperature increase 5–8 °C, i.e. at a rate of 0.00025 °C/year to 0.0004 °C/year. Carbon isotope modelling by McInerney and Wing (2011) suggest the peak rate of carbon addition in the range of 0.3–1.7 GtC year^{-1}, lower by a large factor compared to the rate of carbon emissions in 2018 (10 GtC year^{-1}). Mass balance calculations by these authors suggest CO_2 increase of between ~1200 ppm and ~1,500 ppm, potentially triggered by a release of 4300 GtC from methane clathrates required to generate a Carbon isotope excursion of $-4.6‰$.

According to Cui et al. (2011), the PETM warming was accompanied by a rapid shift in the isotopic signature of sedimentary carbonates, suggesting that the event was triggered by a massive release of carbon to the ocean–atmosphere system, whereas the source, the rate of emission and total amount of carbon involved remain poorly constrained. Simulations suggest that peak rate of carbon addition was probably in the range of <0.4 ppm CO_2 per year. This event led to ocean acidification from pH ~8.2 to ~7.5 and a major extinction of benthic foraminifera, with widespread oxygen deficiency in the ocean as a possible cause (Zachos et al. 2008).

According to Wright and Schaller (2013) carbon and oxygen isotope evidence suggest the PETM was triggered by the release of carbon on the scale of ~3000 GtC as methane over a period of ~6000–7000 years. Zeebe et al. (2016) calculated stable carbon and oxygen isotope records from the onset of the PETM at the New Jersey shelf, finding that carbon release persisted over at least 4000 years, which constrains the PETM carbon release rate to less than 1.1 GtC per year. This is an order of magnitude less than 2014 emissions of 10 GtC per year, surpassing recorded geological analogues. The study concludes that future ecosystem disruptions are likely to exceed the extinctions observed at the PETM.

According to Gehler et al. (2016) the Carbon Isotope Excursion (CIE) resulted from a massive release of carbon into the atmosphere. The authors combine an established oxygen isotope paleo-thermometer with a newly developed triple oxygen isotope paleo-CO_2 barometer. The results are consistent with previous estimates of PETM temperature change and suggest that not only CO_2 but also massive release of seabed methane was the driver for CIE and PETM. The effects of the PETM as represented in sedimentary and fossils isotopic records are portrayed in Figs. 2.5 and 2.6.

The above suggests the Anthropocene rate of CO_2 rise and global warming since 1750 (2–3 ppm/year in the twenty-first century; 0.9 °C/270 year since 1850 = 0.0033 °C/year) is faster by an order of magnitude than all the estimates for the PETM (Table 5.1), underpinning the extreme nature of current global warming.

2.3   Cenozoic Climates

A sharp decline at the end-Eocene (~34 Ma) in mean global CO_2 from ~1120 to 560 ppm (Royer 2006) and of temperatures by more than 5 °C over a period of ~10 Ma (Fig. 2.3), ended with the formation of the Antarctic ice sheet (Zachos et al. 2001). Upper Eocene temperature decline is attributed to CO_2 capture associated with erosion of the rising Himalayan and Alpine mountain chains (Ruddiman 1997). The sharp freeze about 34 Ma is likely

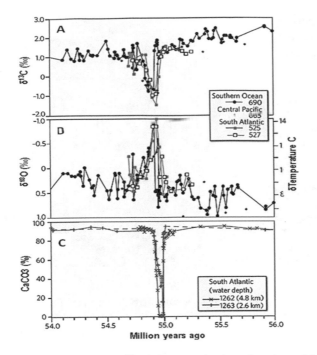

Fig. 2.5 **a** Carbon isotope records $\delta^{13}C$ (Photosynthetic carbon is enriched with ^{12}C); **b** oxygen $\delta^{18}O$ isotope records and temperature for the PETM based on benthic foraminiferal $CaCO_3$ data (the $^{18/16}Oxygen$ ratio decreases with evaporation [higher T]); **c** $CaCO_3$ from drill holes in the South Atlantic. The decrease in $CaCO_3$ reflects lowered pH and increased dissolution of $CaCO_3$. Zachos et al. (2008), Springer/Nature

related to the opening of the Drake Passage between South America and West Antarctica, isolating Antarctica from the influence of northern warmer currents.

Studies of the post-34 Ma glacial–interglacial era based on proxies of temperature (oxygen isotopes, Ca/Mg ratios), greenhouse gas levels (fossil plant pores/stomata), wind (dust content in marine sediments), salinity (boron) and biological relics (pollen, organic remains of algal alkenones, plant wax residues in sediments) allow identification of principal climate change trajectories (Royer et al. 2001; Berner 2004, 2006) (Fig. 2.3).

Glacial states since ~34 Ma were interrupted by protracted warming events accompanied by sea level rises, including in the late Oligocene (~25 Ma), early and mid-Miocene (~17–14 Ma) and late Pliocene (~3.1–2.9 Ma) (Zachos et al. 2001) (Fig. 2.3). On the other hand a possible occurrence of short-lived climate spikes during the Oligocene, Miocene and Pliocene is rendered unlikely by the long-term atmospheric residence time of CO_2, which should have shown in the isotopic record. The suggested CO_2

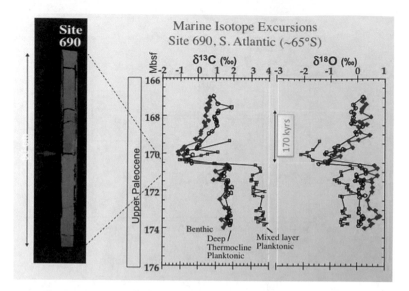

Fig. 2.6 The PETM record in a core located off Antarctica. The red arrow represents extinction of benthic foraminifera. Decreasing carbon isotope values indicate input from greenhouse gases from either organic or volcanic source (Kennett and Stott 1991, NASA)

longevity of up to about 10^4 years for the PETM (Zachos et al. 2008, Eby et al. 2009; Solomon et al. 2009), place constraints on the likelihood of such transient extreme events.

With the exception of possible short-lived temperature spikes, mean greenhouse gas and temperature variations during upper Cainozoic warming events suggest slow greenhouse change rates on the scale of 10^{-3} ppm CO_2 per year and $<10^{-5}$ °C per year (Table 5.1).

The decline of greenhouse gas levels from about ~400 ppm in the Miocene to 180–280 ppm CO_2 cycles following the end-Pliocene about ~2.6 Ma ago has been associated with a decline of mean global temperatures by 2–3 °C and of sea levels by a mean of 22 ± 10 m (Miller et al. 2012). The decline in GHG increased the direct exposure of the Earth surface to variations in insolation controlled by Milankovic cycle orbital cycles (Hansen et al. 2007a, b, 2008, 2011; Hansen and Sato 2012).

Protracted cooling through the Pleistocene glacial cycles is manifested by the growth of the Laurentian, Greenland and Fennoscandian ice sheets, leading to a decline in the intensity of the hydrological cycle and to replacement of rainforests with open savanna, where Hominids evolved (deMenocal 2004). The transition from tropical to savanna environments in Africa, accompanied by faunal changes from tropical to arid zone-type species, including abrupt diversifications of antelopes at ~2.8, 1.8–1.7 and 0.8–0.7 Ma, was associated with an increase in climate variability and an enhanced pace of evolution. Human evolution accords with these transitions in terms of variability selection, diversification and appearance of Olduvan stone tools from ~2.7 Ma and Acheulean stone tools from ~1.7 Ma (Klein and Edgar 2002).

The temporal increase in intensity of the glacial–interglacial polarity and the abrupt nature of glacial terminations, driven by solar pulsations maxima in the order of ~40–60 W m^{-2} (Overpeck et al. 2006) in high northern latitudes, in turn triggered powerful feedback effects (Lashof 2018) from the ice/water albedo flip process (Hansen et al. 2007, 2008; Hudson 2011). Amplifying feedbacks to ice melt included: (a) reduction in albedo (reflection) due to melting of ice and infrared absorption by melt water, extended rock surfaces and spreading vegetation, constituted self-amplifying feedback processes, and (b) reduced CO_2 and organic carbon sequestration in warming oceans (McSweeney 2015). Release of greenhouse gases may have initially lagged behind insolation peak temperatures and the onset of ice/water albedo-flip cycles by approximately 700 years (Hansen et al. 2007). Glacial termination events, preceded by low-variability lulls (Dakos et al. 2008), involved release of greenhouse gases and temperature rise at rates on the order of 10^{-2} ppm CO_2/year and 5.10^{-4} °C/year, i.e. faster than Miocene and Pliocene rates but lower than modern rates (~2–3 ppm/year; +0.0097C/year; Table 5.1).

The fastest rates of climate shifts recorded prior to the nineteenth and twentieth centuries pertain to intra-glacial cycles during ~75–20 kyear-ago, including 6000–7000 years-long Heinrich cooling cycles (Yokoyama and Esat 2011) and ~1470-year-long Dansgaard–Oeschger (D–O) cycles (Ganopolski and Rahmstorf 2002; Braun et al. 2005). The D–O cycles affect melting of the Greenland and Laurentian ice sheets and are also manifested in tropical latitudes (Broecker 2000). Temperature rises in Greenland ice cores at the outset of D–O cycles reaching ~8 °C within a few decades are attributed to solar signals amplified by ocean currents. Braun et al. (2005) modeled the D–O cycles in terms of combination of 90 and 210 years-long solar cycles (Dergachev and Raspopov 2000). Approximate mean global temperature rise rates are estimated in the range of ~0.01–0.2 °C per year, which is of similar to or higher than 20–twenty-first century temperature rise rates (Table 5.1). CO_2 increases associated with D–O cycles are estimated as ~20 ppm, with CO_2 rise rates of ~0.2 ppm per year, an order of magnitude below modern rates (Table 5.1). The D–O cycle represent the most abrupt <1000 years-long

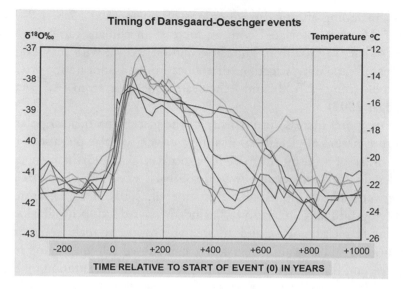

Fig. 2.7 D–O oscillations showing a very abrupt warming phase followed by a slow cooling phase Rahmstorf and Stocker (2004). Springer

oscillations of a similar magnitude to modern global changes related to abrupt insolation cycles rather than changes in greenhouse gas levels (Fig. 2.7).

Abrupt climate shifts including rapid freezing are exemplified by the onset of stadial cooling episodes following temperature peaks, for example the 12.9–11.7 kyear Younger Dryas cold phase in Greenland where sharp temperature oscillations occurred over periods as short as 1–3 years (Steffensen et al. 2008). The Younger Dryas stadial was followed at ~8.2 kyear by a sharp temperature decline of several degrees Celsius in the North Atlantic region, associated with discharge of cold water from the Laurentian ice sheet through Lake Agassiz (Wiersma et al. 2011). Both the Younger Dryas and the 8.2 kyear Laurentian stadial represent phases typical of upper Pleistocene interglacials (Cortese et al. 2007) (Fig. 2.8). A decline in CO_2 and methane since the Holocene Optimum (~8.0 kyear ago) was subsequently replaced by a slow rise of CO_2 from about ~6000 BC and of methane from ~4000 BC, likely related to farming and to fires.

According to Ruddiman (2003) and Kutzbach et al. (2010) the natural interglacial cycle has been interrupted by Neolithic land clearing and burning, halting a decline in CO_2 and methane and thereby an onset of a post-Holocene glacial. Other authors regard the rise in greenhouse gases from the mid-Holocene as a natural perturbation in the interglacial, comparable to the 420–405 kyear Holsteinian interglacial (Broecker and Stocker 2000). The late Holocene is interrupted by mild regional warming phase about ~900–1400

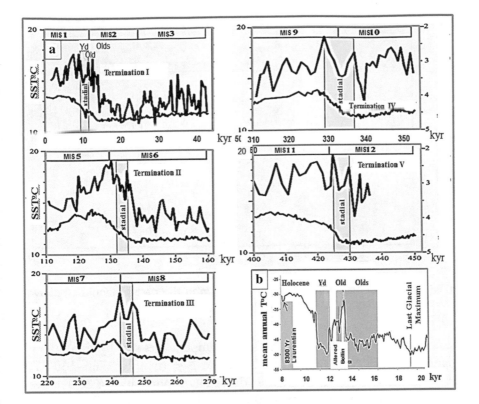

Fig. 2.8 **a–f** Evolution of sea surface temperatures in 5 glacial-interglacial transitions recorded in ODP-1089 at the sub-Antarctic Atlantic Ocean. Grey lines—d^{18}O measured on Cibicidoides plankton; Black lines—sea surface temperature. Marine isotope stage numbers are indicated on top of diagrams. Note the grey zone-marked stadials following interglacial peak temperatures, analogous to the Younger dryas (12.9–11.7 kyear) preceding the onset of the Holocene; **b** The Last Glalcial Termination. Modified after Cortese et al. (2007). With permission

AD (Medieval Warm Period) and a cool phase at ~1650–1600 AD (Little Ice Age—LIA), broadly corresponding to periods of solar insolation.

According to Ruddiman (2003) orbital-scale timing indicates ice sheets should have appeared 6000–3500 years ago and CO_2 and CH_4 levels should have fallen steadily from 11,000 years ago until the present. However new ice did not appear, and CO_2 and CH_4 began to increase at 8000 and 5000 years ago by 40 ppm and 250 ppb, respectively, implicating anthropogenic emissions from land clearing, fires, farming and animal husbandry Roberts (1998). By 2000 years ago complex agricultural practices had replaced natural forests in large areas of southern and western Europe, India, and eastern China. Early deforestation in these regions is more than large enough to account for the observed increase in CO_2 since 8000 years ago. During the

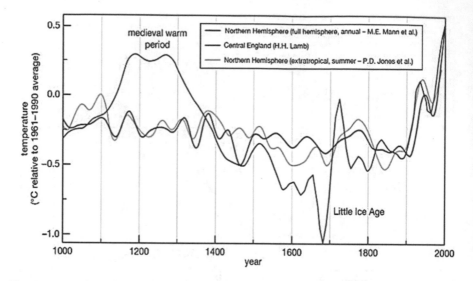

Fig. 2.9 Northern hemisphere temperatures during the past millennium relative to 1000–2000 AD (Mann et al. 1999). With permission

Medieval Warm Period (~900–1400 AD) temperatures rose by about 0.25 °C (Mann et al. 1999). During the Little Ice Age (mainly 1700–1600 AD) temperatures plunged across the Northern Hemisphere by 0.6 °C relative to the average temperature between 2000 and 1000 AD (Fig. 2.9).

3

Human Origins (Abbreviated from Groves 2016)

Ancestors of mammals, referred to as mammal-like reptiles, separated from reptiles and birds in the Carboniferous and evolved during the Permian, surviving in part during the Permian–Triassic mass extinction (Kump et al. 2005) becoming abundant in the Jurassic and the Cretaceous. Small mammals, distinguished from mammal-like reptiles by their middle-ear and dental structure, survived the dinosaurs and other giant reptiles and underwent diversification during the Tertiary and the Pleistocene. Living mammals are divided into Prototheria and Theria, where Prototheria are represented by the platypus and echidna confined to Australia and New Guinea. Paleocene monotreme fossils are known from South America and differ from other mammals in laying eggs and secreting milk for the young. From molecular clock calculations humans separated from the chimpanzees about 6–7 million years ago, followed by re-combination about 4 million years ago. The species *Sahelanthropus tchadensis* is regarded as the earliest member of the Hominin (Figs. 3.1 and 3.2). Other relatives are gorillas, orangutans and gibbons. Following the K–T Mass Extinction placental mammals underwent an adaptive radiation filling the ecological niches left vacant by extinct land animals.

The human lineage may have separated from that of chimpanzees some 6–7 million years ago, but about 4 million years ago there was an episode of interbreeding (Cartmill and Smith 2009; Stringer and Andrews 2011). The oldest identified biped Hominin species are the 6 million years-old *Sahelanthropus tchadensis* discovered in Chad, *Orrorin tugenensis* found in Kenya

A. Y. Glikson, *The Event Horizon: Homo Prometheus and the Climate Catastrophe*, https://doi.org/10.1007/978-3-030-54734-9_3

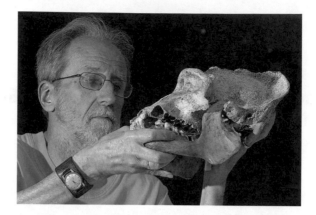

Fig. 3.1 Colin Peter Groves (by permission of Phyl Dance)

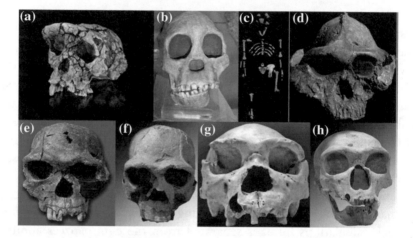

Fig. 3.2 Cranium of **a** *Sahelanthropus.tchadensis* (https://en.wikipedia.org/wiki/File): Sahelanthropus_tchadensis_TM_266-01-060-1.jpg); **b** Taung Child (https://en.wikipe dia.org/wiki/Taung_Child); **c** skeleton of *Australopithecus afarensis* (Lucy) (https://en.wikipedia.org/wiki/File:Lucy_blackbg.Jpg); **d** cranium of *Paranthropus boisei*-Nairobi (https://en.wikipedia.org/wiki/File:Paranthropus.boisei-Nairobi.jpg); **e** cranium of *Homo habilis* (https://en.wikipedia.org/wiki/Homo_habilis); **f** cranium of *Homo ergaster* (https://en.wikipedia.org/wiki/Homo_ergaster); **g** cranium of *Homo heidelbergensis*; **h** cranium of *Homo sapiens-neanderthalis* (Wikipedia/Creative Commons)

and somewhat younger *Ardipithecus ramidus* from Ethiopia. Younger species are represented by approximately 3.9–4.2 Ma years old "*Australopithecus anamensis*" found in northern Kenya, and the 3.75–3.00 Ma *Australopithecus afarensis* from Ethiopia, Kenya and Tanzania, with a cranial volume of 350–500 cm^3, living in a range of environments (Su and Harrison 2015) from arid regions dotted with shrubs to well-watered forested areas drained by seasonal streams.

The species *Australopithecus Africanus* is recorded in the Sterkfontein region in South Africa between 2.0 and 2.5 Ma, parallel with the genus *Paranthropus* specializing in consumptions of heavy vegetation, coexisting with early Homo in East African sites (Koobi Fora, Omo, Olduvai) as well as in the South African caves of Swartkrans and Kromdraai, all dated between 2 and 1 Ma. Ancestral species to Homo may be represented by the 2.5 Ma *Australopithecus garhi* and *Australopithecus sediba*. Lake Turkana, northern Kenya, evidence exists for marked lake level fluctuations related to climate changes contemporaneous with human existence, indicating its ability to cope with extreme climatic variability (Maslin et al. 2014). The appearance of *Homo rudolfensis* (750 cm^3) at 2.058 ± 0.034 Ma, *Homo habilis* (>1.4 Ma) and *Homo ergaster* (800–900 cm^3) is regarded as a long-distance bipedal walker ancestor. These species lived during a period when open links existed between Lake Turkana and the Indian Ocean.

The earliest equivalents in Asia are represented by 1.75 Ma *Homo georgicus* from Dmanisi, Georgia, intermediate between *Homo habilis* and *Homo ergaster*. A near-equivalent species in Java, Homo erectus, is dated as 1.6 Ma-old, persisting in Java until less than 500 kyear ago. The earliest human fossil from China from a group named Homo pekinensis is 1.63 Ma-old. Descendants of *Homo ergaster* spread to Europe and Asia about 600 kyear as the larger-brained species *Homo heidelbergensis*. The European/West Asian branch evolved into a stocky, big-brained *Homo neanderthalensis* and the African branch into *Homo sapiens*. About 60,000 years ago the species migrated from Africa via Sinai or southern Arabia into tropical Asia and Australia, then northwest into Europe, and towards the end of the Pleistocene into the Americas.

Molecular clock calculations indicate human and chimpanzee lineages separated some 6 million years-ago, diverged from a common ancestor. Comparative morphology and DNA analysis agree chimpanzees are human's closest living relatives. Further relatives are gorillas, then orangutans, then gibbons, Old World monkeys and so on. Matsuzawa (2012) showed that the working memory of chimpanzees is strikingly superior to humans, which may be connected to their need for instant decision-making in the natural habitat. However, while apes do plan, it is for only the short-term future, for example bringing two different types of sticks to a termite mound, a sturdy one to dig into the mound and a long, flexible one to insert afterwards in order to extract termites to eat, or to retrieve food before others can discover it. By contrast human ontogeny is geared towards manipulation and learning, especially social learning. Syntax appears to be uniquely human and the openness of language and its infinite flexibility specifically human in its fullest extent, but the rudiments of this it seems can be grasped by chimpanzees (Watson et al. 2015).

Humans are notoriously interested in aesthetic activities, art, music, poetry and story-telling practiced in all societies. It would seem that, like tool-making and language, the predisposition for art may already have been there in the human lineage when it separated from the chimpanzee lineage. In the archaeological record this is expressed in stone or cave paintings. The earliest archaeological records of art may be in the Middle Pleistocene. Some of the oldest art comes from Blombos Cave in South Africa, dating to 70 kyear, followed by several sites in Europe dating to perhaps up to 40 kyear, including the 33 kyear Chauvet Cave painting, and in some instances by *Homo neanderthalensis*.

4

Fire and Human Intelligence

Strained wire-like fingers strum a guitar
White strings of pain take you afar
Blue visions, a tall angel by the gate
Leads down steps, where fate awaits
A door swings, brilliant light engulfs
A huge hall, walls' cryptic hieroglyphs
Bright ray beams flash, piercing toward
A radiant core, mortal's reward
Where the eternal cosmic flame
Creates the life it shall reclaim.

Originating around pre-historic camp fires, fire has become a core of human beliefs, intrinsically associated with the emergence of nature spirits and of gods in the human mind and in religious and cultural traditions. Fire induces a major change in entropy[1] in nature and in human nature. Human respiration dissipates 2–10 cal per minute, a camp fire covering one square meter releases approximately 180,000 calories per minute, and the output of a 1000 MW/h power plant expends some 2.4 billion calories per minute, namely some 500 million times the mean energy level of individual human respiration, magnifying human power by many orders of magnitude.

---

[1]A thermodynamic quantity representing the unavailability of a system's thermal energy for conversion into mechanical work, often interpreted as the degree of disorder or randomness in the system. "the second law of thermodynamics says that entropy always increases with time".

Plato thought the gods created humans from earth and fire, where the mastery of fire explained human dominance on the basis of pyro-technology, the key for the evolution of human intelligence and beliefs (Pyne 2016). Around 450 BC Aristotle believed the world was made of four elements: earth, water, air, and fire, an idea which became the cornerstone of philosophy, science, and medicine for two thousand years. Darwin regarded language and fire the two most significant achievements of humanity. The language of prehistoric humans however may not have been far advanced over other primates or even the chemical signals and responses exchanged among the colonial arthropods through the use of pheromones.

According to Malouf (1996) Prometheus had a foresight and could think ahead, but Epimetheus could only reflect later. It was the stealing of the fire from the gods for which Prometheus, metaphorically, had to endure punishment through eternal torture, as do humans due to repeated wars and plagues through the ages. But it is around fires that pre-historic people developed their concept of the gods, admired through a complex system of beliefs and rituals.

In the view of Lewis Mumford (1972): *"My point of departure in analyzing technology, social change and human development, concerns the nature of man. And to begin with I reject the lingering anthropological notion, first suggested by Benjamin Franklin and Thomas Carlyle, that man can be identified, mainly if not solely, as a tool-using or tool-making animal: Homo faber. Even Henri Bergson, a philosopher whose insights into organic change I respect, so described him. Of course man is a tool-making, utensil-shaping, machine-fabricating, environment-prospecting, technologically ingenious animal—at least that! But man is also—and quite as fundamentally—a dream-haunted, ritual-enacting, symbol-creating, speech-uttering, language-elaborating, self-organizing, institution-conserving, myth-driven, love-making, god-seeking being, and his technical achievements would have remained stunted if all these other autonomous attributes had not been highly developed. Man himself, not his extraneous technological facilities, is the central fact. Contrary to Mesopotamian legend, the gods did not invent man simply to take over the unwelcome load of disagreeable servile labor."*

In itself the use of stone tools (Biro et al. 2013) can hardly explain the emergence of *Homo sapiens*, as many other species use elaborate construction methods, including the colonial insects and some birds and mammals, for example the Kapuchin monkeys who use stone tools.[2] Nor can visual art provide a fundamental difference between humans and animals. Thus

---

[2] https://www.sciencenews.org/article/capuchin-monkey-stone-tool-use-evolution-3000-years.

Bowerbirds decorate their nests to attract a mate[3] and other animals may express themselves through art.[4] It would appear in this respect the difference between humans and other species is more in terms of degree, rather than in terms of kind. This raises the question whether a criterion exists which fundamentally distinguishes humans from other species and which can account for their extraordinary development of their faculties?

Wrangham (2009) attributed the increase in brain size and the drop in tooth size at 1.9–1.7 Ma in *Homo ergaster*, relative to *Homo habilis* to onset of cooking of meat, enhancing the brain blood supply. Easier digestion of raw vegetable improved the calorie intake and relieved early humans from energy-consuming chewing. Cooking breaks down the connective tissues in meat and softens the cell walls of plants to release their stores of starch and fat (Adler 2013).[5] The calories to fuel the bigger brains of successive species of hominids lead to the tall narrow-waist body of *Homo sapiens*. Since their mastery of fire hominins grew taller and leaner, shedding much of their original hair cover, allowing perspiration, cooling and thereby long range chase and hunt of animals.

Furthermore, cooking frees time allowing humans to spend more leisure time around fires. According to Adler (2013) *"Wherever humans have gone in the world, they have carried with them two things, language and fire. As they traveled through tropical forests they hoarded the precious embers of old fires and sheltered them from downpours. When they settled the barren Arctic, they took with them the memory of fire, and recreated it in stoneware vessels filled with animal fat."*

Inherent in ancient fire mythologies is the illegitimate nature of the acquisition of fire by humans, symbolized by Prometheus (Cartwright 2013), the Titan who, breathing life into human clay figures, stole the fire from the gods and gave it to the human beings he had created. The prophetic Prometheus mythology of fire stolen from the gods is coming to haunt humanity (Fig. 4.1). Once clearing and burning of the forests evolved into combustion of every available molecule of carbon derived from excavated coal, drilled and extracted oil and gas, large parts of Earth have become poisoned by emanations from the toxic remnants of past biospheres.

According to the Rig Veda the hero Mātariśvan recovered fire which had been hidden from mankind. In Cherokee myth, after Possum and Buzzard had failed to steal fire, grandmother Spider used her web to sneak into the

---

[3] https://www.theguardian.com/science/neurophilosophy/2012/jan/19/1.

[4] https://en.wikipedia.org/wiki/Animal-made_art; https://www.wired.com/2012/02/animal-art/; https://cheezburger.com/4033029/11-artistic-animals-that-express-themselves-through-art.

[5] https://www.smithsonianmag.com/science-nature/why-fire-makes-us-human-72989884/.

Fig. 4.1 A black-figure Lakonian kylix, c. 570–560 BCE, depicting the Titan Atlas carrying the world on his shoulders and Prometheus being tormented by an eagle sent by Zeus to eat his liver as punishment for giving mankind the gift of fire, stolen from Hephaistos. (Gregoriano Etrusco Museum, Vatican). A metaphor for humankind's suffering. Wikipedia commons

land of light and stole fire, hiding it in a clay pot. Among various Native American tribes of the Pacific Northwest and First Nations, fire was stolen and given to humans by Coyote, Beaver or Dog. According to some Yukon First Nations people, Crow stole fire from a volcano in the middle of the water. According to the Creek Indians, Rabbit stole fire from the Weasels. In Algonquin myth, Rabbit stole fire from an old man and his two daughters. In Ojibwa myth, Nanabozho the hare stole fire and gave it to humans. In Polynesian myth, Maui stole fire from the Mudhens. In the Book of Enoch, the fallen angels and Azazel teach early mankind to use tools and fire.

According to Goudsblom (1995) people have become truly human by learning to domesticate fire and cook food, leading to the birth of agriculture and the industrial revolution and representing huge leaps forward and the transformation from pre-historic culture to civilization. In his book *Catching Fire: How cooking made us human* Wrangham (2009) points out humans are the only animals that cook their food and that *Homo erectus* emerged about two million years ago as a result of this unique trait. Twomey (2011) inferred a high degree of human cognition and cooperation required for fire use and control. This included technologies of igniting, preserving and transporting fire, planning of cooking requiring collection of firewood, precautionary measures of avoiding burn and fire-spreading.

Over the millennia pre-historic humans, spending much of their lives around fires, were mesmerized and inspired by the flames, fire becoming an integral part of their life and faith. As evidenced by burial rites associated with ashes it is likely around fires that the concept of the gods originally emerged. To keep camp fires going humans needed to undertake long-term planning, including collection of firewood in advance and protection of fire from wind, rain and storms. These activities demand cognitive behaviour associated with accessing, maintaining and using fire. The mastery of fire allowed the genus Homo to increase entropy in nature by orders of magnitude. The ever-changing flickering life-like flames, reflecting human dynamics, coupled with the emanating radiative heat, inevitably engender emotional effects, hopes, dreams, premonitions and fears. Fire has become a focus of human beliefs.

In his comprehensive and eloquent treatise on fire Pyne (2016) wrote "*The possession of fire was unique, which humans knew at their origins. More than anything else, fire defined them and segregated them from the rest of creation; myths that depict the origin of fire account equally for the origins of humans because the latter depended on the former. Typically, the proto-humans are help-less. Typically, some culture hero—a daring animal, a Titan, a cunning youth, a pitying god—steals fire from a potentate who hoards it as an expression, if not the source, of his own power. With fire, humans begin to act for themselves.*" and "*Repeated fire was, paradoxically, a means of perpetual renewal. It was a simple matter to decide that a similar logic governed the cosmos, that the world might begin and end with fire or enjoy immortality by passing through fire-induced cycles of death and rebirth.*"

In an attempt to appease the Gods civilizations developed rituals of human sacrifice in war and on altars, monuments to the great hunters and conquerors (Fig. 4.2) and self-declared gods (Fig. 4.3). It is the fear of death humans tried to conquer through acts of great courage, dangerous adventures, wars and a belief in an afterlife, defying death through death.

Inherent in reverence of nature, the rocks, plants and animals, is a connection of human Earthlings with their mother Earth. The transformation from pantheism to cults of Olympian gods and subsequently to Monotheism, commencing with dominant god and reaching a peak early in the first Millennium with the rise of Christianity, constituted a fundamental cultural shift leading to a view of Earth as but a corridor to heaven. At the final stage, a space cult aspires for human conquest of the planets while the sixth extinction of species is in progress (Barnosky et al. 2012).

Fig. 4.2 Nebamun hunting birds in the marshes using cats, fragment of a scene from the tomb-chapel of Nebamun, Thebes, Egypt Late 18th Dynasty, around 1350 BC. Wikipedia Public Domain

An atavistic child sacrifice aimed at appeasing the gods, as exercised by the Canaanites, Olmec, Aztecs, Maya, Inca and other cultures, constitutes a historic background for sending the young to war, making space for the privileged, the generational sacrifice of 37 million lives in World War I being a more recent example. With the exception of invasions aimed at feeding purposes among the arthropods it is hard to think of any species which conducts massive regular mass killings as humans do. The natural kindness of individual humans is swamped by murderous rituals, culminating in a repeated to perpetual state of war.

A primeval hunting and killing cult inherent in the male macho, reflected in the modern world by devastation of nature, deforestation, pollution and climate change, is contrasted with the dominant life-giving, nurturing and cultivation nature of women. With exceptions, matrilineal societies such as exist in parts of Africa, Southeast Asia, and India are relatively more peaceful and nature-friendly as compared to patriarchal societies. The Minangkabau of Sumatra, Indonesia, is the world's largest matrilineal society where land and houses are inherited through female lineage.

Fig. 4.3 Colossal statue of the god Nabu, eighth century BCE. From Nimrud, Mesopotamia. The Iraq Museum

As expressed by an Aztec war song: (1300–1521 AD):

There is nothing like death in war,
Nothing like flowering death,
So precious to him who gives life,
Far off I see it, my heart yearns for it.

The killing by Achilles of Penthesilea, queen of the Amazons (Fig. 4.4), signifying the victory of male hunters over child-bearing women and the triumph of patriarchal over matriarchal societies, symbolizes the rise of killing and war in human history.

The smelting of metals, commencing with copper and tin in the Bronze Age, allowed societies to use fire for the crafting of farming tools and of weapons of war, constituting the precursors of combustion of fossil fuels. Smelting spread through the Middle East, where the Assyrian Empire spanned the Early Bronze Age to the late Iron Age (2500–605 BC). It

Fig. 4.4 Achilles kills Penthesilea, queen of the Amazons, a tribe of warrior women believed to live in Asia Minor, their main concern in life was war, likely in defense from marauding males

is during the Bronze Age about 3150 BC that the Egyptian civilization flourished, attaining the highest architectural and artistic achievements. Independently the Moche civilization of South America 100–700 AD) and the Incas of Peru and the Calchaquí people of Northwest Argentina developed bronze technology. The harnessing of fossil fuels, coal, petroleum and gas, the toxic high energy remnants of past biospheres, allowed the species to do both, alter large parts of the Earth and write its own death warrant.

5

The Age of Consequences

5.1 The Anthropocoene Hyperthermal

Compared to the glacial-interglacial cycles with an amplitude of ~3.0 W/m² and 4–5 °C, during the Anthropocene greenhouse gas forcing has risen by more than 2.0 W/m², equivalent to more than >2 °C, above the peak Holocene interglacial, which constitutes an abrupt event over not much longer than a lifetime (Fig. 5.1).

As the climate stabilized, from about 7000 years ago, the application by Neolithic civilizations of iron tools, production of excess grain and animal husbandry allowed human creativity, imagination, dreams, aggression, fear of death and worship of nature and of the gods to be expressed through the construction of monuments for immortality, death rituals and genocidal wars in the name of superior powers. Further to cutting and burning trees in a war against the forests, since the onset of the industrial age combustion and extraction of carbon from fossil biospheres set the stage for an anthropogenic oxidation event leading to an abrupt shift in state of the atmosphere–ocean–cryosphere system. The consequent progressive mass extinction of species is tracking towards levels commensurate with those of the past five great mass extinctions of species, constituting a geological event horizon in the history of planet Earth.

According to Ruddiman (2003), Ruddiman (2018) and Kutzbach et al. (2010) greenhouse-gas emissions from early agriculture before 1850 were large enough to alter atmospheric composition and global climate substantially. Ruddiman (2018) states *"Fifteen years after publication of Ruddiman*

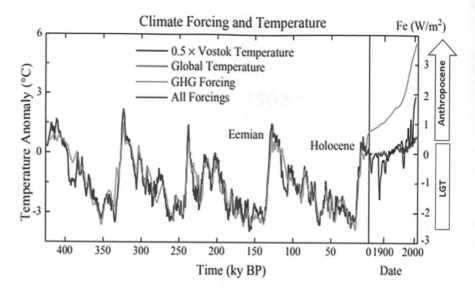

Fig. 5.1 Glacial-interglacial temperatures and GHG forcing for the last 420,000 years based on the Vostok ice core, with the time scale expanded for the Anthropocoene. The ratio of temperature and forcing scales is 1.5 °C per 1 W/m². The temperature scale gives the expected equilibrium response to GHG change including slow feedback surface albedo change. Modern forcings include human-made aerosols, volcanic aerosols and solar irradiance (Hansen et al. 2008). Courtesy James Hansen

(2003), the early anthropogenic hypothesis is still debated, with relevant evidence from many disciplines continuing to emerge. Recent findings summarized here lend support to the claim that greenhouse-gas emissions from early agriculture (before 1850) were large enough to alter atmospheric composition and global climate substantially".

Beginning with the eighteenth century, combustion of fossil fuels has led to the release of a total of near to 250 GtC or in another estimate 340 GtC (billion ton of carbon). The mean post-1880 temperature rise rate (~0.0067 °C/year) and the mean post-1960 rate (0.0139 °C/year) exceed those of the PETM which involved more than 5 °C of warming in 15–20 thousand years, i.e. ~0.0003–0.00075 °C/year, closer to the rate of the last glacial termination [LGT] (~0.00039 °C/year) and the Eemian interglacial (~0.0004 °C/year). The industrial CO_2 rise rates exceed those estimated for the K–T and PETM events by an order of magnitude (Table 5.1).

- K–T: ~350–500 ppm to at least ~2300 ppm within about 10,000 years, 0.18 ppm/year (Beerling et al. 2002)
- PETM: raising CO_2 from ~1000 ppm to ~1700 ppm over 4000 years, i.e. 0.175 ppm/year (Zeebe et al. 2009).

Table 5.1 Comparison of mean global temperature and CO$_2$ rise rates during the Cenozoic, including the ~66 Ma K–T impact events (Beerling et al. 2002), the 55.9 Ma PETM hyperthermal event (Zachos et al. 2008; Zeebe et al. 2009), end-Eocene freeze and formation of the Antarctic ice sheet (34–32 Ma) (Zachos et al. 2001), Oligocene (Zachos et al. 2001; Liu et al. 2009; Pekar and Christie-Blick 2007), Miocene (Kurschner et al. 2008) and end-Pliocene (Zachos et al. 2001; Beerling and Royer 2011; de Conto and Pollard 2003; Denton 2010) thermal rises, glacial terminations (Hansen et al. 2007), Dansgaard-Oeschger cycles (Ganopolski and Rahmstorf 2002; Jouzel et al. 2007), 8.2 kyear event (Wagner et al. 2002) intra-Holocene events (IPCC 2007) and Anthropocene climate change (IPCC 2007; NOAA 2019)

Age	Interval (warming period)	Mean land and sea temp change C	Warming rate (C/year)	CO$_2$ change (ppm)	CO$_2$ change rate (ppm/year)	References
K–T impact 64.98 Ma	10,000 years	Short freeze followed by ~+7.5 C	~0.00075	~500–2300 ppm	0.18 ppm/year	Beerling et al. (2002)
PETM 55.9 Ma	~6000 to 7000 years	~+5–9 C	~0.0008–0.0015	~835–1500 ppm Added 665 ppm	~0.095–0.11 ppm/year	Cui et al. (2011), Zeebe et al. (2009)
PETM	20,000 years	~5–8 C	0.00025–0.0004	Added 1200–1500 ppm	0.06–0.075 ppm/year	McInerney and Wing (2011)
Eocene–Oligocene freeze 34.2–34.0	34–33.5 Ma	~5.4 C Over 0.5 Ma		~1120–560 ppm Over ~10 Ma		Liu et al. (2009), DeConto and Pollard (2003)

(continued)

Table 5.1 (continued)

Age	Interval (warming period)	Mean land and sea temp change C	Warming rate (C/year)	CO_2 change (ppm)	CO_2 change rate (ppm/year)	References
End-Oligocene Warming ~24.7	~200,000 years	~+4 C	0.00002	500–900 ppm	0.002	Pekar and Christie-Blick, (2007)
Mid-Mmmiocene 20–18 Ma	~200,000 years	~+1.5 C	0.000007	~300–520 ppm	0.0011	Kurschner et al. (2008)
End-pliocene	4–3 Ma	~+1 C	0.000001	~250–400 ppm	0.00015	Zachos et al. (2001), Beerling and Royer (2011)
Eemian—glacial termination	11,000 years	+~5 C	0.0004	+100 ppm	0.009	Hansen et al. (2007), Petit et al. (1999), EPICA (2004)
Dansgaard-Oeschger—21 cycles of ~1500 years each	~75–15 kyear	~6–8 C	0.01–0.2	+20 ppm	0.2	Ganopolski and Rahmstorf (2002), Jouzel et al. (2007)

Age	Interval (warming period)	Mean land and sea temp change C	Warming rate (C/year)	CO_2 change (ppm)	CO_2 change rate (ppm/year)	References
Younger dryas Interglacial stadial	12.9–11.7 kyear	~−15 C in GISP2 ice core		−7 ppm		
8.2 kyear stadial	~100 years	−3.3 C in the North Atlantic		−25 ppm in ~300 years	−0.08	Wagner et al. (2002)
Medieval warm period (MWP)	~400 years	~0.4–0.5 C	~0.001	5 ppm	~0.012	IPCC (2007) Chapter 4
Little ice age (LIA)	~60 years	~−0.4 C	~−0.006	−5 ppm		IPCC (2007) Chapter 4
Post-1750	269 years	+0.9 C	+0.0033	280–408 ppm	~0.48 ppm/year	IPCC (2007)
Post-1975	44 years	+0.84 C	+0.0097	330–408 ppm	~1.73–2.86 ppm/year	NOAA (2019)

The current growth rate of atmospheric greenhouse gases, in particular over the last 70 years or so, is the fastest recorded in the Cenozoic since the Cretaceous boundary (K–T) impact event 66.4 million years-ago and the Paleocene-Eocene Thermal Maximum (PETM) 55.9 million years-ago, constituting an extreme event in the recorded history of Earth. Beginning with the eighteenth century, the combustion of fossil fuels has led to the release of 910 billion tons of carbon dioxide ($GtCO_2$) or of 248 billion ton carbon (GtC) by human activity. An estimate by the World Meteorological Organization (WHO) suggests emission of 375 GtC and 1374 $GtCO_2$. Consequently currently CO_2 concentration in the atmosphere has reached about 410.27 ppm (November 2019), as compared to the 280–300 ppm range prior to the onset of the industrial age. By the early-twenty-first century, at the current CO_2 rise rate of 2–3 ppm/year, the fastest rise rate recorded in the Cenozoic, the CO_2-equivalent level which combines the radiative role of CO_2, CH_4 and N_2O, has reached near 500 ppm-CO_2-e (Figs. 5.2 and 5.3).

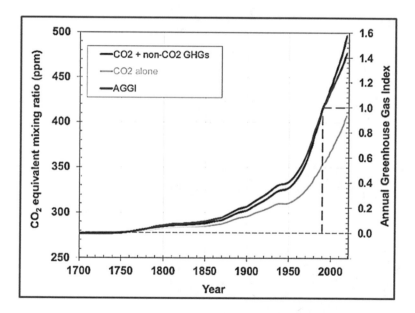

Fig. 5.2 Growth of CO_2-equivalent level and the annual greenhouse gas Index (AGGI (The index uses 1990 as a baseline year with a value of 1. The index increased every year since 1979. https://www.co2.earth/annual-ghg-index-aggi)) (Butler and Montzka 2019). Measurements of CO_2 to the 1950s are from (Keeling et al. 2008) and from air trapped in ice and snow above glaciers. Equivalent CO_2 amounts (in ppm) are derived from the relationship between CO_2 concentrations and radiative forcing from all long-lived greenhouse gases. Public domain

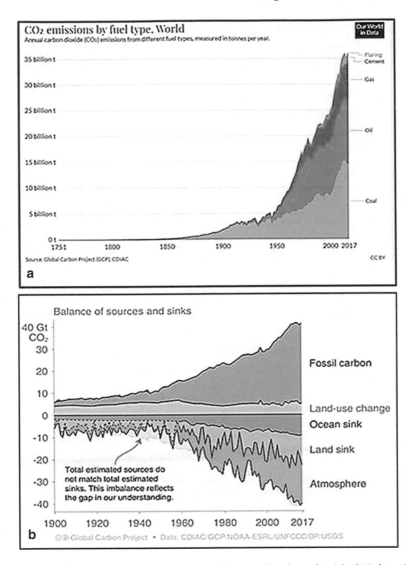

Fig. 5.3 **a** CO₂ emissions by fuel type, World Annual carbon dioxide (CO₂) emissions from different fuel types, measured in tonnes per year; **b** Combined components of the global carbon budget as a function of time, for CO₂ emissions (grey) and emissions from land-use change (brown), as well as their partitioning among the atmosphere (blue), ocean (turquoise), and land (green). Ritchie and Roser (2019) Global Carbon Budget (2018). https://www.icos-cp.eu/GCP/2018

Direct evidence for changing climate patterns is provided by the expansion of the tropics (Lucas et al. 2014) and migration of climate zones toward the poles, estimated at a rate of approximately 56–111 km per decade. As these dry subtropical zones shift, droughts will worsen and overall less rain will fall in most warm temperate regions. Poleward shifts in the average tracks of tropical and extratropical cyclones are already happening. This is likely to continue as the tropics expand further. As extratropical cyclones move, they shift rain away from temperate regions that historically rely upon winter rainfalls for their agriculture and water security. Australia is highly vulnerable to expanding tropics as about 60% of the continent lies north of 30°S (Turton 2017).

A *"tipping point"* in the climate system is a threshold that, when exceeded, can lead to large changes in the state of the system, or where the confluence of individual factors combine into a single stream. Hansen et al. (2008) introduced the term *"tipping element"* to describe subcontinental-scale subsystems of the Earth system that are susceptible to being forced into a new state by small perturbations. A report by the National Academy Press 2011 states: *"As the planet continues to warm, it may be approaching a critical climate threshold beyond which rapid (decadal-scale) and potentially catastrophic changes may occur that are not anticipated—because of complex feedback dynamics and existing computational limitations—by climate models that are tuned to modern conditions."*

In so far as a tipping point can be identified in current developments of the climate system (Lenton et al. 2008) the weakening of the Arctic boundary (Baek-Min Kim et al. 2014) indicated by increased undulation of the jet stream and the intrusion through the boundary of southward-shifting cold fronts and northward-migrating warm air masses, herald a possible tipping point.

Global greenhouse gases have reached a level exceeding the stability threshold of the Greenland and Antarctic ice sheets, which are melting at an accelerated rate (Rignot et al. 2019). The current extreme rise rates of GHG (2.86 ppm CO_2/year) and temperature (0.15–0.20 °C per decade since 1975) raise doubt with regard to future linear and curvilinear climate projections. Allowing for the transient albedo-enhancing effects of sulphur dioxide and other aerosols, mean global temperature has reached approximately 2.0 degrees Celsius above pre-industrial temperatures (Hansen et al. 2011). Current greenhouse gas forcing and global mean temperatures are approaching Miocene Optimum-like composition, bar hysteresis effects of reduced ice sheets.

Low-lying land areas, including coral islands, delta and low coastal and river valleys would be flooded due to sea level rise to Miocene-like sea levels of approximately 40 ± 15 m (Betzler et al. 2018) above pre-industrial levels.

Accelerated flow of ice melt water flow from ice sheets into the oceans (Hansen et al. 2016) is reducing temperatures over tracts in the North Atlantic and circum-Antarctic oceans (Rahmstorf et al. 2015; Glikson 2019). Strong temperature contrasts between cold polar-derived fronts and warm tropical-derived air masses lead to extreme weather events, retarding habitats, in particular over coastal regions. As partial melting of the large ice sheets proceeds the Earth's climate zones continue to shift polar-ward (Kidston 2012; Environmental Migration Portal 2015) reducing polar ice sheets and temperate climate zones and expanding tropical to super-tropical regions such as existed in the Miocene (5.3–23 million years ago).

In so far as a climate sensitivity of 3 ± 1.5 °C for doubling of CO_2 is assumed (IPCC AR5), a large part of which is masked by direct and indirect effects of emitted sulfur aerosols (Hansen et al. 2011), the mean global temperature rise since 1880 may have reached significantly higher values than +0.9 °C, consistent with an estimate as ~1.5 °C for continental regions by the Berkeley Earth (2018). Temperatures are amplified in the Polar Regions to +2.3 °C, mainly due to the effects of albedo flip process and the expanding areas of open sea water.

Since 1975–1976 rise in carbon emissions and reduced SO_2 aerosols associated with clean-air policies were expressed by a marked upward trend of the mean global temperature, reached 408 ppm, exceeding any in the late Cenozoic. The rise in atmospheric CO_2 by ~2.0–3.0 ppm/year, driving temperature rise rates of ~0.015 °C/year during 1959–2015, and of ~0.02 °C/year during 2010–2019, exceeds all recorded rates, excepting regional Atlantic Ocean Dansgaard-Oeschger cycles of 0.01–0.2 °C/year (Table 5.1, Fig. 2.7).

Emissions of both CO_2 and SO_2 have peaked during WWI and WWII, reflecting the growing military-industrial industry, the wars and the post-war economic boom accelerating from about 1950 (CDIAC 2010). A lull in the rise of mean global temperature, accompanied with a mild degree of cooling during 1940–1975, related to rising levels of sulfur aerosols and a low in the 11-years sun spot cycle, was terminated by the abrupt resumption of warming from 1975 when atmospheric CO_2 level climbed to ~330 ppm and aerosol levels declined due to clear air policies.

Hansen et al. (2008) regard a CO_2 level of ~350 ppm as the maximum level allowable before amplified feedbacks lead to tipping points beyond human control. Following the 1998 peak El Nino event, mean temperature rise rates declined relative to the 1975–1998, likely related to a surge in SO_2 emissions (Smith et al. 2011) and a decline in sunspot activity. Peak temperature anomalies unprecedented in the instrumental record were reached in 2010 (+0.7 °C), 2014 (+0.74 °C) and 2019 (+0.96 °C) relative to 1880.

Positive feedbacks to global warming include summer exposure of open water in polar latitudes, replacing the reflective properties of ice surfaces with absorption by dark water. Hudson (2011) estimates the rise in radiative forcing due to total removal of Arctic summer sea ice as ~0.7 W m^{-2}, close to the current Earth radiative imbalance of 0.58 ± 0.15 Wm^{-2} (Hansen et al. 2011), in part offset by increased summer cloudiness. The subsequent increase in evaporation leads to the advance of cold vapor-laden air masses into the sub-Arctic and the North Atlantic, perpetrating snow storms, crossing the weakened Arctic boundary, and colliding with warm air masses. Positive feedbacks occur due to the release of methane from Arctic permafrost, where reserves are estimated at ~900 GtC, and of clathrate from the Arctic Ocean. Other vulnerable methane accumulations include high-latitude peat lands (~400 GtC) and tropical peat lands (~100 GtC) (Canadell P, 2009).

Significant reserves of CH_4 are held in the Arctic seabed where the release of CH_4 to the overlying ocean and subsequently to the atmosphere has been believed to be restricted by impermeable subsea permafrost, which has sealed the upper sediment layers for thousands of years (Shakhova et al. 2019). In the regions where permafrost exists, hydrate-bearing sediment deposits can reach a thickness of 400–800 m. Shallow hydrate deposits are predicted to occupy ~57% of the East Siberian Arctic Shelf (ESAS) seabed. It has been suggested that destabilization of shelf Arctic hydrates could lead to large-scale enhancement of aqueous CH_4. More recently, according to Shakhova et al. (2019), the Russian scientist in charge of an East Siberian Arctic Ocean expedition, a methane burst from the East Siberian Shelf could happen at any time and needs only a trigger.

Since the 1980s, global temperature, ice melt rates and sea level rise have been lagging behind the rise in atmospheric radiative forcing, masked by sulfur aerosols (Hansen et al. 2012). With the migration of climate zones toward the poles (Fig. 5.5), ensuing greening occurs in some high-latitude zones (Lucht et al. 2002) and some desert areas such as the southern Sahara Desert. On the other hand increased desertification of temperate zones occurs in other regions, including northwest China, North Africa, southern Europe, south and southwest Australia and southern Africa, where the forests become affected by heat waves and firestorms. Much of the greening is dominated by grasses, whose photosynthetic activity and carbon storage capacity are approximately an order of magnitude less than trees. Warming of the oceans leads to a decrease in CO_2 solubility, lowered pH, and thereby a decrease in biological calcification and CO_2 sequestration. Increased evaporation enhances the hydrological cycle, including abrupt precipitation events and floods, a pole-ward shift in the storm tracks (Bengtsson et al. 2006), intensification of cyclones and associated disruption of vegetation (Rahmstorf and Coumou 2011; Hansen et al. 2012).

Current trends are shifting the atmosphere to a state analogous to the end-Pliocene, before 2.6 Ma ago, a period when large parts of the Greenland and West Antarctic ice sheets did not exist. Hansen et al. (2016) suggest ice mass loss from the most vulnerable ice, sufficient to raise sea level several meters, is better approximated as an exponential rather than linear response. Doubling times of ice melt rates every 10, 20 or 40 years yield multi-meter sea level rise exceeding the contribution to sea level rise from thermal expansion and mountain glaciers.

In Antarctica mass loss increased from 40 ± 9 Gt/year in 1979–1990 to 50 ± 14 Gt/year in 1989–2000, 166 ± 18 Gt/year in 1999–2009, and 252 ± 26 Gt/year in 2009–2017 (Rignot et al. 2019). In Greenland decadal mass loss increased six-fold since the 1980s (Mouginot et al. 2019). In July 2019 warm air from Europe's heat wave set temperature records melting at Summit Station. About 90% of the ice sheet surface melt runoff was estimated at 55 billion tons, about 40 billion tons more than the 1981–2010 average for the same time period, leading to more extensive bare ice and flooded snow areas (NSIDC 2019). Antarctica has lost some 2.71 trillion tons of ice since 1994, mostly from West Antarctic. Parts of eastern Antarctica are already affected by large scale ice melt, including the largest glacier—the Totten Glacier. All this hardly taken into account by the powers that be, as they keep talking about the "future" mainly in monetary terms, arming themselves to the teeth with weapons of mass destruction.

A rise of mean global temperatures toward +4 °C (PIK 2012) would track toward greenhouse Earth conditions such as existed during the early Eocene some 50 million years ago (Zachos et al. 2001). Sea level rise represents the sum-total manifestation of global temperature changes, including thermal expansion of water, melting ice sheets and mountain glaciers, albeit lagging behind radiative forcing levels. Foster and Rohling (2013), on the basis of temperature–ice volume–sea level relations for five periods during the last 40 million years, point to the dominant role of CO_2 in controlling ice volumes and sea levels. According to these authors, a CO_2 level stabilized at 400–450 ppm is likely to lead in the long term to a 9 m sea level rise. The sea level rise/temperature rise ratios of the glacial terminations (SL/T ~ 7–20) exceed those of 20–twenty-first centuries' ratios (SL/T < 1.0) by more than an order of magnitude, giving a measure of the lag of sea level rise behind mean global temperatures (Glikson 2016).

Based on paleoclimate studies, the current levels of CO_2 of 408 ppm measured at Mauna Loa, 2019, and of CO_2-equivalent of near 500 ppm (Fig. 5.1), a value that includes the effects of methane and nitrous oxide, commit the atmosphere to a warming trend tracking toward the upper stability level of 500 ± 50 ppm CO_2 of the Antarctic ice sheet (Zachos et al. 2001).

Comparisons between temperature rise rates (ΔT/year) indicate the following (Table 5.1):

1. Modern temperature rise rates (0.0097 °C/year) exceed those of glacial terminations (Eemian ~0.0004 °C/year) by more than an order of magnitude;
2. Since the start of the satellite sea level record in 1993, the average rate of sea level has been about 3.1 mm per year.

This SL rise rate is similar or somewhat lower than those of glacial terminations (~3–10 mm/year), and both are significantly lower than Dansgaard-Oeschger sea level rise rates (10 < mm/year). Equilibrium between sea level rise and the potential temperature rise of 2–3 °C above pre-industrial levels implies Pliocene-like sea levels of 25 ± 12 m above Holocene levels (Chandler et al. 2008), although due to ice sheet hysteresis the time lag of sea level rise is unclear.

An onset of an irreversible change in state of the global climate system, referred to as *tipping points* (Lenton et al. 2008), occurs when the climate system shifts to a state at which amplifying positive feedbacks drive climate change further until negative feedbacks, such as a depletion in the source of greenhouse gases or cooling due to decrease in solar insolation stabilize a new state (Charbit et al. 2008). A tipping point may result from a synergy of multiple processes, including melting and collapse of the large ice sheets, melting of permafrost, boreal forest dieback, tundra loss, Indian and west African monsoon shifts, Amazon forest dieback, large-scale fires and changes to the ENSO circulation and ocean deep water formation patterns. Transient negative feedbacks ensue from the flow of large amounts of cold ice melt water into the oceans (Glikson 2019). According to Hansen et al. (2012) *'Burning all fossil fuels would create a different planet than the one that humanity knows. The palaeoclimate record and ongoing climate change make it clear that the climate system would be pushed beyond tipping points, setting in motion irreversible changes, including ice sheet disintegration with a continually adjusting shoreline, extermination of a substantial fraction of species on the planet, and increasingly devastating regional climate extremes'.* Whereas the precise nature and timing of a global tipping point remains unclear, several of the criteria indicated by Lenton et al. (2008) are on track.

According to Ganopolski et al. (2016), a temperature decline toward an ice age has been on track and was narrowly missed before the outset of the industrial age in the eighteenth century. Instead, anthropogenic CO_2 emissions and low orbital eccentricity of the Earth are likely to postpone the next glacial inception by at least 100,000 years. The 0.2–1.0 °C rise in sea surface temperature during 1800–2011 (Hansen et al. 2011, 2013) has intensified the hydrological evaporation/precipitation cycle, ensuing in cyclones and floods. The rise in land temperatures has led to heat waves and fires, raising the frequency of extreme weather events. A rise in the intensity of cyclones in the Atlantic Ocean (Trenberth and Shea 2006) is not globally uniform. An increase in frequency and/or intensity of extreme weather events is indicated by the IPCC (IPCC-AR5, 2012) and NOAA (GFDL 2019), which states: "*Models project substantial warming in temperature extremes by the end of the twenty-first century. It is virtually certain that increases in the frequency and magnitude of warm daily temperature extremes and decreases in cold extremes will occur in the twenty-first century on the global scale. It is very likely that the length, frequency and/or intensity of warm spells, or heat waves, will increase over most land areas. Based on the A1B and A2 emissions scenarios, a 1-in-20 year hottest day is likely to become a 1-in-2 year event by the end of the twenty-first century in most regions, except in the high latitudes of the Northern Hemisphere, where it is likely to become a 1-in-5 year event*".

An analysis of the relations between long-term climate trends and the incidence of extreme weather events (Rahmstorf and Coumou 2011) finds that the number of record-breaking heat events increases approximately in proportion to the ratio of warming trend to short-term standard deviation or variability. A longer term climatic warming trend increases the number of heat extremes whereas short-term variability tends to decrease it. Projections of twenty-first century climate trends suggest that temperatures and sea levels are lagging behind equilibrium conditions, in part due to aerosol solar-shielding effects and hysteresis of the ice sheets. Further rise in temperatures would shift the climate toward conditions such as existed during the Pliocene (~5.3–2.6 Ma) and the peak Miocene (~16 Ma), when temperatures were ~2–3 °C higher than the pre-industrial Holocene (Chandler et al. 2008).

Transient cooling periods may ensue by analogy to stadial conditions such as that are represented by the Younger Dryas (12.9–11.7 kyear; Steffensen et al. 2008) and the 8.5 kyear events (Wagner et al. 2002) as a consequence of the growth in the cold ice–melt water region south of Greenland (Rahmstorf et al. 2015) and around Antarctica (Glikson 2019). A collapse of the

Atlantic Mid-Ocean Circulation (AMOC) would result in major climate cooling effect, in particular in northern Europe and North America. Based on temperature reconstruction for the (AMOC) a weakened circulation after 1975 is an unprecedented event in the past millennium (Rahmstorf et al. 2015). The Anthropocene climate state has already reached Miocene-like conditions, when atmospheric CO_2 levels exceeded > 400 ppm (Londono et al. 2018) and the combined CO_2-equiivalent reached near 500 ppm (Fig. 5.2). Whereas the future role of emissions may be self-limiting due to a decline of industry, amplifying feedbacks from land and ocean may continue to push CO_2 and temperatures upwards.

A key question regarding current global warming is concerned with the adaptability and survivability of flora and fauna species at high rates of environmental and climatic changes. Ceballos et al. (2015) indicate the current rate of mammal and vertebrate extinctions is up to 100 times higher than the background rate of 2 mammal extinctions per 10,000 species per 100 years, indicating that a sixth mass extinction is under way. Quintero and Wiens (2013) studied 17 time-calibrated phylogenies of major tetrapod clades and climatic data from distributions of > 500 extant species in relation to climate change between 2000 and 2100. Their results indicate that adaptation of species to projected changes through the twenty-first century would require adaptation rates unprecedented among vertebrate species. Thomas et al. 2004, projecting species' distributions relative to future climate scenarios, assessed extinction risks for about 20 percent of the Earth's terrestrial surface. On the basis of mid-range climate-warming scenarios for 2050, the study predicts 15–37 percent of species in the relevant regions are committed to extinction. Minimal climate-warming scenarios produce ~18% of species committed to extinction, mid-range warming models produce ~24%, and maximum change produces ~35% of species committed to extinction.

5.2   Accelerated Melting of the Ice Sheets

The shift in state of the Earth's climate is most acutely manifested in the Polar Regions, where warming is driven by the ice-water albedo flip, opening dark sea-water to insolation, replacing the highly reflecting ice and snow. Warming decreases the temperature contrast between the Arctic and sub-polar regions, leading to weakening of the jet stream boundaries (Fig. 5.9). Consequently the boundaries are breached by outflow of cold air fronts, such as the recent "Beast from the East" event, as well as penetration of warm air masses. The fast rate of the Anthropocoene temperature rise, as compared to the LGT and PETM (Fig. 5.7) has major effects on the capacity of flora and fauna for adaptation to the new conditions. As the poles keep warming, to date by a mean of ~2.3 °C, the rate of shrinking of the ice sheets has accelerated by a factor of more than six fold.

The threshold of collapse of the Greenland ice sheet, retarded by hyster esis,[1] is estimated in the range of 400–560 ppm CO_2, already transgressed at the current 496 ppm CO_2-equivalent (Fig. 5.8). The Greenland mass loss increased from 41 ± 17 Gt/yr in 1990–2000, to 187 ± 17 Gt/year in 2000–2010, to 286 ± 20 Gt/year in 2010–2018, or six fold since the 1980s, or 80 ± 6 Gt/year per decade, on average.

The greenhouse gas level and temperature conditions under which the East Antarctic ice sheet formed during the late Eocene 45–34 million years ago are estimated as ~800–2000 ppm and up to 4 degrees Celsius above pre-industrial values, whereas the threshold of collapse is estimated as 600 ppm CO_2. The total mass loss from the Antarctic ice sheet increased from 40 ± 9 Gt/year in 1979–1990 to 50 ± 14 Gt/year in 1989–2000, 166 ± 18 Gt/year in 1999–2009, and 252 ± 26 Gt/year in 2009–2017. Based on satellite gravity data the East Antarctic ice sheet is beginning to breakdown in places (Jones 2019). According to this author *"East Antarctica is the coldest spot on earth, long thought to be untouched by warming. But now the glaciers and ice shelves in this frigid region are showing signs of melting, a development that portends dramatic rises in sea levels this century and beyond."* Notably the Totten Glacier (Rignot et al. 2019), which could be irreversible. According to Levermann and Mengel 2014 the Wilkes Basin in East Antarctica contains enough ice to raise global sea levels by 3–4 m (Fig. 5.4).

5.3   Migration of Climate Zones

The expansion of tropical zones and the polar-ward migration of subtropical and temperate climate zones are leading to a change in state in the global climate pattern. The migration of arid subtropical zones, such as the Sahara (Fig. 5.5a), Kalahari (Fig. 5.5b) and central Australian deserts (Fig. 5.5c) at the expense of temperate climate zones ensues in large scale droughts, such as in inland Australia and southern Africa. In the northern hemisphere expansion of the Sahara desert northward is manifested by heat waves and fires across the Mediterranean and Europe (Fig. 5.5a).

---

[1]where a physical property lags behind changes in the effect causing it.

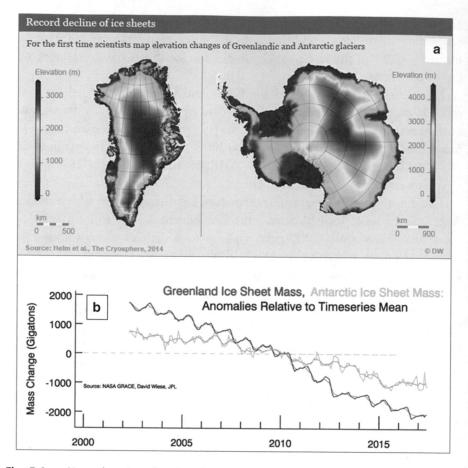

Fig. 5.4 **a** New elevation showing the Greenland and Antarctic current state of the ice sheets accurate to a few meters in height, with elevation changes indicating melting at record pace, losing some 500 km³ of ice per-year into the oceans; **b** Ice anomaly relative to the 2002–2016 mean for the Greenland ice sheet (magenta) and Antarctic ice sheet (cyan). World Meteorological Organization. Data are from GRACE; **c** the melting of Arctic sea ice 1990–2017, National Snow and Ice Data Centre (NCIDC). These maps show sea ice age of late March 1990 (left) and 2016 (right), around the time of the winter maximum. Younger, thinner ice appears in shades of blue; older, thicker ice appears in shades of pale green and white. Ice-free ocean water is dark gray, and land areas are light gray. Image courtesy NOAA

As the globe warms, to date by a mean of near ~1.5 °C, or ~2.0 °C when the masking effects of sulphur dioxide and other aerosols are considered, and by a mean of ~2.3 °C in the Polar Regions, the expansion of warm tropical latitudes and the polar-ward migration of climate zones ensue in large scale droughts in subtropical latitudes, such as in inland Australia and southern Africa. A similar trend is taking place in the northern hemisphere

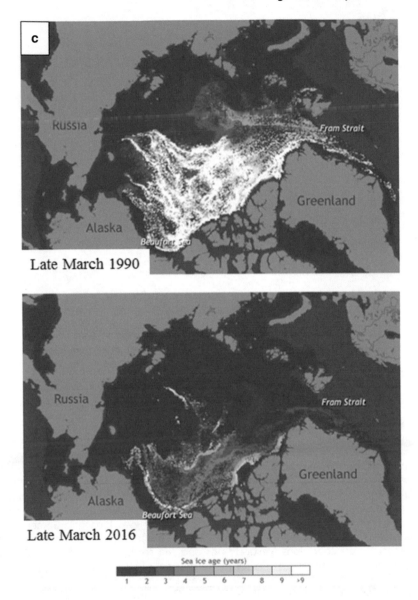

Fig. 5.4 (continued)

where the Sahara desert is expanding northward, with consequent heat waves across the Mediterranean and Europe. Since 1979 the planet's tropics have been expanding polar-ward by in both hemispheres. A leading commentator called this Earth's bulging waistline. Future climate projections suggest this expansion is likely to continue, driven largely by emissions of greenhouse gases and black carbon, as well as warming in the lower atmosphere and the

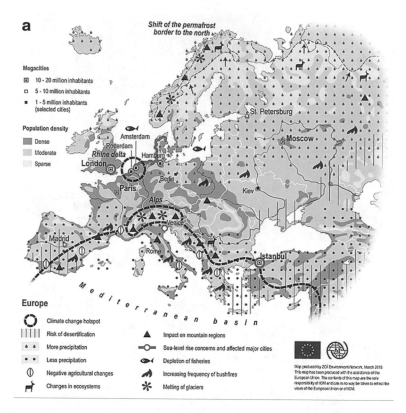

Fig. 5.5 a Migration of the subtropical Sahara climate zone (red spots) northward into the Mediterranean climate zone leads to warming, drying and fires over extensive parts of Spain, Portugal, southern France, Italy, Greece and Turkey, and to melting of glaciers in the Alps. Migration, Environment and Climate Change, International Organization for Migration, Geneva, Switzerland. With permission of the IOM UN Migration. The regional impacts of climate change map extracted from The Atlas of Environmental Migration (Ionesco D., Mokhnacheva D. and Gemenne F., Routledge, Abingdon, 2017), p. 63 © IOM (Mokhnacheva, Ionesco), Gemenne, Zoï Environment Network, 2015. Sources: IPCC (2013, 2014); **b** Southward encroachment of Kalahari Desert conditions (vertical lines and red spots) leading to warming and drying of parts of southern Africa. With permission of the IOM UN Migration. The regional impacts of climate change map extracted from The Atlas of Environmental Migration (Ionesco D., Mokhnacheva D. and Gemenne F., Routledge, Abingdon, 2017), p. 63 © IOM (Mokhnacheva, Ionesco), Gemenne, Zoï Environment Network, 2015. Sources: IPCC (2013, 2014); **c** Drying parts of southern Australia, including Western Australia, South Australia and parts of the eastern States, accompanied by increasing bushfires. With permission of the IOM UN Migration. The regional impacts of climate change map extracted from The Atlas of Environmental Migration (Ionesco D., Mokhnacheva D. and Gemenne F., Routledge, Abingdon, 2017), p. 63 © IOM (Mokhnacheva, Ionesco), Gemenne, Zoï Environment Network, 2015. Sources: IPCC (2013, 2014)

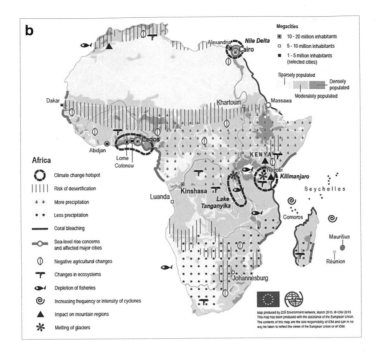

Fig. 5.5 (continued)

oceans. This expansion is associated with heating and drying at the expense of originally temperate habitats rich in flora and fauna.

5.4 Climate Extremes

Since the bulk of extant terrestrial vegetation has evolved under glacial-interglacial climate conditions, where GHG range between 180 - 300 ppm CO_2, global warming is turning large parts of Earth into a tinderbox, ignited by natural and human agents (Bowman 2009) (Fig. 5.6). By July and August 2019, as fires rage across large territories, including the Amazon forest, dubbed the *Planet's lungs* as it enriches the atmosphere in oxygen. When burnt the rainforest becomes a source of a large amount of CO_2, with some 72,843 fires in Brazil this year and extensive bushfires through Siberia, Alaska, Greenland, southern Europe, parts of Australia and elsewhere, the planet's biosphere is progressively transformed. As reported by Barlow and Lees (2019).

According to *Prof. Hans Joachim Schellnhuber "Climate change is now reaching the end-game, where very soon humanity must choose between taking*

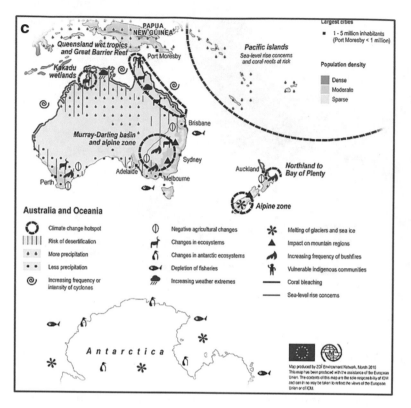

Fig. 5.5 (continued)

unprecedented action or accepting that it has been left too late and bear the consequences".

In the words of James Hansen (in Morton 2018) *"We've reached a point where we have a crisis, an emergency, but people don't know that … There's a big gap between what's understood about global warming by the scientific community and what is known by the public and policymakers".* According to the Potsdam Institute for Climate Impact Research (PIK) *"Human kind's emissions of greenhouse gases are breaking new records every year. Hence we're on a path towards 4-degree global warming probably as soon as by the end of this century. This would mean a world of risks beyond the experience of our civilization—including heat waves, especially in the tropics, a sea-level rise affecting hundreds of millions of people, and regional yield failures impacting global food security".*

As the Earth warms the frequency and intensity of extreme weather events are increasing, including heat waves, fires, hurricanes and floods, signifying rising temperatures. This is associated with the migration of climate zones toward the poles, including expansion of the tropics, a shift of temperate

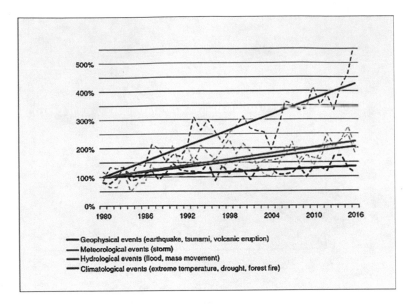

Fig. 5.6 20–21st Century extreme weather events. Natural catastrophes to 2018. Credit MunichRe NatCatSERVICE

climate zones toward the poles and weakening of the polar jet stream boundary. Given a 20-twenty-first centuries warming rate faster than the last major thermal event, linear warming trajectories such as are projected by the IPCC may reach punctuated tipping points, transient reversals and stadial cooling episodes.

According to Hansen et al. 2008 the rise in radiative forcing during the Last Glacial Termination (LGT—18,000–11,000 years BP), associated with amplifying feedbacks, has driven GHG radiative forcing by approximately ~3.0 Wm^{-2} and a mean global temperature rise of ~4.5 °C (Fig. 5.1) or, i.e. of similar order as the Anthropocene rise since about 1900. However the latter has been reached within a time frame at least 30 times shorter than the LGT, underpinning the extreme nature of current global warming.

According to NOAA GHG forcing in 2018 has reached 3.101 Wm^{-2} relative to 1750 (CO_2 = 2.044 Wm^{-2}; CH_4 = 0.512 Wm^{-2}; N_2O = 0.199 Wm^{-2}; CFCs = 0.219 Wm^{-2}) with a CO_2-equivalent of 492 ppm (Fig. 5.2). The rise in GHG forcing during the Anthropocene since about 1800 AD, intensifying since 1900 AD and sharply accelerating since about 1975, has induced a mean of ~1.5 °C over the continents above pre-industrial temperature, or >2.0 °C when the masking role of aerosols is discounted, implying further warming is in store.

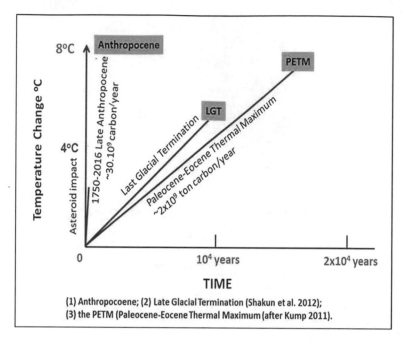

Fig. 5.7 A comparison between rates of mean global temperature rise during: (1) the last Glacial Termination (after Shakun et al. 2012); (2) the PETM (Paleocene-Eocene Thermal Maximum, after Kump 2011); (3) the late Anthropocene (1750–2019); and (4) an asteroid impact. In the latter instance temperature due to CO_2 rise would lag by some weeks or months behind aerosol-induced cooling

During 1750–2019 greenhouse gas levels have risen from ~280 ppm to above > 408 ppm and to 496 ppm CO_2-equivalent (Fig. 5.2), the increase of CO_2 reaching near-46 percent above the original atmospheric concentration. However linear climate change projections are rare in the recent climate and linear warming trajectories such as are projected by the IPCC may overlook amplifying feedbacks from land and oceans, punctuated tipping points, transient reversals and stadial events.

5.5   An Uncharted Climate Territory

Elements of the Quaternary glacial-interglacial history are relevant to an understanding of current and future developments with the climate. The rise of total greenhouse gas (GHG), expressed as CO_2-equivalent, to 496 ppm CO_2-e (Fig. 5.2), within less than a century represents an extreme event in the atmosphere, raising GHG and temperatures from the Holocene range to the Miocene (34–23 Ma) range, when CO_2 level was between 300 and

530 ppm, within a century or two, a geological blink of the eye. As the glacial sheets disintegrate, cold ice-melt water flowing into the ocean form large pools of cold water, currently manifested by the growth of cold regions in north Atlantic Ocean south of Greenland and in the Southern Ocean fringing Antarctica (Fig. 5.8), by analogy to stadial events in the wake of peak interglacial phases over the last 450,000 years.

Golledge et al. (2019) show meltwater from Greenland will lead to substantial slowing of the Atlantic overturning circulation, while meltwater

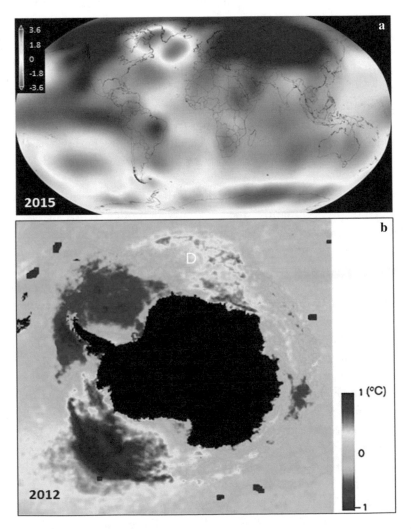

Fig. 5.8 a Global warming map (NASA 2018). Note the cool ocean regions south of Greenland and along the Antarctic. Credits: Scientific Visualization Studio/Goddard Space Flight Center; **b** 2012 Ocean temperatures around Antarctica (NASA 2012)

from Antarctica will trap warm water below the sea surface, increasing Antarctic ice loss. Bronselaer et al. (2018) suggest Antarctic meltwater would increases the formation of Antarctic sea ice, associated with warming of the subsurface ocean around the Antarctic coast. This would delay temperature rise of 1.5 and 2 degrees Celsius by more than a decade, cause enhanced drying of the Southern Hemisphere and reduce drying of the Northern Hemisphere. The meltwater-induced subsurface ocean warming could lead to further ice-sheet and ice-shelf melting through a positive feedback mechanism, highlighting the importance of including meltwater effects in simulations of future climate.

Whereas the effect of low-density ice melt water on the surrounding oceans is generally not included in many models, depending on amplifying feedbacks, prolonged Greenland and Antarctic melting and consequent cooling of surrounding ocean sectors, as well as penetration of freezing air masses through weakened polar boundaries, may have profound effect on future climate change trajectories. According to Hansen et al. (2007) warming according to the IPCC 'Business As Usual' (BAU) scenario *"would lead to a disastrous multi-meter sea level rise on the century timescale"*, delayed by hysteresis.

Climate projections for 2100–2300 by the IPCC-AR5 Synthesis Report, 2014 portray predominantly linear to curved model trends of greenhouse gas, global temperatures and sea level changes. These models appear to take limited account of amplifying feedbacks from land and ocean and of the effects of the flow of cold ice-melt into the oceans.

Zachos et al. (2008) state: *"If fossil-fuel emissions continue unabated, in less than 300 years pCO_2 will reach about 1,800 ppmv, a level not present on Earth for roughly 50 million years. Both the magnitude and the rate of rise complicate the goal of accurately forecasting how the climate will respond …" By the year 2400, it is predicted that humans will have released about 5,000 GtC to the atmosphere since the start of the industrial revolution if fossil-fuel emissions continue unabated and carbon-sequestration efforts remain at current levels. This anthropogenic carbon input, predominantly CO_2 would eventually return to the geosphere through the deposition of calcium carbonate and organic matter. Over the coming millennium, however, most would accumulate in the atmosphere and ocean. Even if only 60% accumulated in the atmosphere, the partial pressure of CO_2 would rise to 1,800 parts per million by volume (p.p.m.v.). A greater portion entering the ocean would decrease the atmospheric burden but with a consequence: significantly lower pH and carbonate ion concentrations of ocean surface layers".*

McInerney and Wing 2011 suggest carbon input during the Paleocene-Eocene Thermal Maximum of more than 2000 GtC into the atmosphere, lasting <20,000 years. Zeebe et al. (2009) suggests an initial input of 3000 GtC from methane clathrates, raising CO_2 from a pre-PETM ~1000 ppm to ~1700 ppm over 4000 years, i.e. 0.175 ppm/year. These values constitute an order of magnitude less than the Anthropocene CO_2 rise, which has reached 2–3 ppm/year.

The rise from pre-industrial levels of ~280 300 ppm represents the sharpest atmospheric greenhouse gas spike recorded since the K–T boundary impact extinction at 66.043 ± 0.011 Ma, with implications for the adaptability and survivability of terrestrial and marine habitats. Last time CO_2 levels were as high as >400 ppm was in the late Pliocene, before 2.6 Million years-ago, when mean temperature was 2–3 °C higher than during the pre-industrial era and sea level was 25 ± 12 m higher than today (NASA, Lindsay 2019). According to this source if global energy demand continues to grow and be met mostly with fossil fuels, atmospheric carbon dioxide will likely exceed 900 ppm by the end of this century.

According to Steffen et al. (2018) "self-reinforcing feedbacks could push the Earth System toward a planetary threshold" and "would lead to a much higher global average temperature than any interglacial in the past 1.2 million years and to sea levels significantly higher than at any time in the Holocene".

Positive feedbacks of global warming include:

A. The albedo-flip of melting sea ice and ice sheets and the increase of the water surface. Hudson (2011) estimates a rise in radiative forcing due to removal of Arctic summer sea ice as 0.7 W/m^2, a value close to the total of methane release since 1750.
B. Reduced ocean CO_2 intake due to decreased solubility of the gas with higher temperatures.
C. Vegetation desiccation and burning in some regions and thereby released CO_2 and reduced evaporation and its cooling effect, weakening the hydrologic cycle through reduced evapotranspiration. This factor and the increase of precipitation in other regions lead to differential feedbacks from vegetation as the globe warms (Notaro et al. 2007).
D. An increase in wildfires, releasing greenhouse gases (Fig. 5.6).
E. Release of methane from permafrost, bogs and sediments.

Linear temperature models appear to take limited account of the negative feedback effects on the oceans of ice melt water derived from the large ice sheets, including the possibility of a significant stadial event such as has been modelled by Hansen et al. (2016) and commenced in oceanic tracts fringing

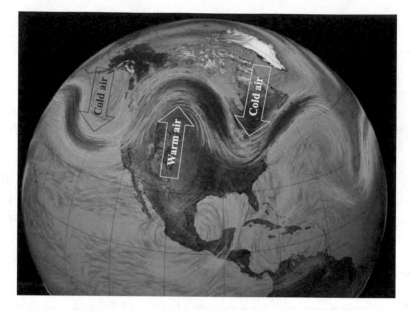

Fig. 5.9 The Arctic jet stream in the summer of 1988. NASA. Extreme melting in Greenland's ice sheet is linked to warm air delivered by the wandering jet stream, a fast-moving belt of westerly winds created by the convergence of cold air masses descending from the Arctic and rising warm air masses from the tropics that flow through the lower layers of the atmosphere. Deep troughs and steep ridges emerge as the denser cold air sinks and deflect warm air regions northward. The pattern propagates across the mid-latitudes as pockets of cold air sporadically creep down from the Arctic, creating contrasting waves and flows that accelerate eastward due to Earth's rotation (Freeman 2019)

Greenland and Antarctica (Fig. 5.10). In the shorter to medium term sea level rises would ensue from the Greenland ice sheet (potentially 6–7 m sea level rise) and West Antarctic ice sheet melt (potentially 4.8 m sea level rise). Referring to major past stadial events, including the 8200 years-old Laurentian melt and the 12.7–11.9 *younger dryas* event, a protracted breakdown of parts of the Antarctic ice sheet could result in major sea level rise and thereby extensive cooling of southern latitudes and beyond, parallel with warming of tropical and mid-latitudes (Fig. 5.10) (Hansen et al. 2016). The temperature contrast between polar-derived cold fronts and tropical air masses is bound to lead to extreme weather events, echoed among other in *Storms of my grandchildren* (Hansen 2010).

The effect of high atmospheric greenhouse gas levels would delay the next ice age by tens of thousands of years (Berger and Loutre, 2002) during which chaotic tropical to hyper-tropical conditions including extreme weather events would persist over much of the Earth until atmospheric CO_2

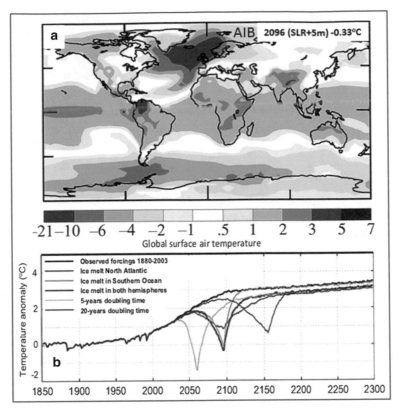

Fig. 5.10 **a** Model surface-air temperature (°C) for 2096 relative to 1880–1920 (Hansen et al. 2016). The projection betrays major cooling of the North Atlantic Ocean, cooling of the circum-Antarctic Ocean and further warming in the tropics, subtropics and the interior of continents; **b** Modeled surface-air temperatures (°C) to 2300 AD relative to 1880–1920 for several ice melt rate scenarios, displaying a stadial cooling event at a time dependent on the ice melt doubling time. Courtesy Prof. James Hansen

and insolation subsided to below ~300 ppm. Humans are likely to survive in relatively favorable parts of Earth, such as sub-polar regions and sheltered mountain valleys, where cooler conditions would allow fauna to remain. An abrupt reduction in carbon emissions is essential. However, since the high level of CO_2-e is activating amplifying feedbacks from land and ocean, global attempts to draw-down of 50–100 ppm of CO_2 from the atmosphere using every effective negative emissions is necessary, including streaming air through basalt and serpentine, biochar, sea weed sequestration, reforestation, sodium hydroxide pipe systems and other methods.

Within and beyond 2100–2300 projections (Fig. 5.10) lies an uncharted climate territory, where further warming of the tropics and subtropics, ongoing melting of the Antarctic ice sheet, further cooling of neighboring sectors of the oceans and clash between warm and cold air masses (Fig. 5.10) ensue in chaotic climate disruptions. Given the thousands to tens of thousands years longevity of atmospheric greenhouse gases (Solomon et al. 2009; Eby et al. 2009), the onset of the next ice age is likely to be delayed on the scale of tens of thousands of years (Berger and Loutre 2002) through an exceptionally long interglacial period (Fig. 5.11). These authors state: "*The present day CO_2 concentration (now >410 ppm) is already well above typical interglacial values of ~290 ppmv. This study models increases to up to 750 ppmv over the next 200 years, returning to natural levels by 1000 years. The results suggest that, under very small insolation variations, there is a threshold value of CO_2 above which the Greenland Ice Sheet disappears. The climate system may take 50,000 years to assimilate the impacts of human activities during the early third millennium. In this case, an "irreversible greenhouse effect" could become the most likely future climate. If the Greenland and west Antarctic Ice Sheets disappear completely, then today's "Anthropocene" may only be a transition between the Quaternary and the next geological period.*"

As conveyed by leading scientists "*Climate change is now reaching the end-game, where very soon humanity must choose between taking unprecedented*

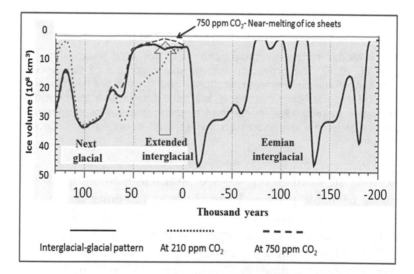

Fig. 5.11 Simulated Northern Hemisphere ice volume (increasing downward) for the period 200,000 years BP to 130,000 years in the future. Time is negative in the past and positive in the future. For the future, three CO_2 scenarios were used: last glacial-interglacial values (solid line), a human-induced concentration of 750 ppm (dashed line), and a constant concentration of 210 ppm inducing a return to a glacial state (dotted line). Modified after Berger and Loutre 2002

action or accepting that it has been left too late and bear the consequences" (Hans Joachim Schellnhuber) … "We've reached a point where we have a crisis, an emergency, but people don't know that … There's a big gap between what's understood about global warming by the scientific community and what is known by the public and policymakers" (James Hansen).

Climate scientists find themselves in a quandary, not dissimilar to medical doctors committed to help the ill but required to communicate grave diagnoses. How do scientists tell people the current spate of extreme weather events, including cyclones, devastating islands from the Caribbean to the Philippines, floods devastating coastal regions and river valleys from Mozambique to Kerala, Pakistan and Townsville, and fires burning extensive tracts of the living world can only intensify in a rapidly warming world? How do scientists tell the people that their children are growing into a world where survival under mean temperatures higher than +2 degrees Celsius above pre-industrial temperatures is likely to be painful and, in some parts of the world impossible, let alone under +4 degrees Celsius projected by the IPCC?

Extensive cyclones, floods, droughts, heat waves and fires increasingly ravage large tracts of Earth. Despite its foundation in the basic laws of physics (the black body radiation laws of Planck, Kirchhoff and Stefan Boltzmann), as well as empirical observations around the world by major climate research bodies (NOAA, NASA, NSIDC, IPCC, World Meteorological Organization, Hadley-Met, Tindale, Potsdam, BOM, CSIRO and others), the anthropogenic origin, scale and pace of climate change remain subject to extensively propagated untruths.

6

Inferno

A huge flare lights an evening
Above grey banks of clouds
As magpies keep on singing
Their music lifts my shrouds.
I stand alone and stare
At god's eternal flame
blinded by the glare
That's uttering a name.
The autumn may yet cede
To signs of a late spring
A mortal can't succeed
Unless they spread a wing
The glow of divine beauty
Stirs red blood in the vein
A will to live, a calling
From root's secret domain
The evening star rose high
To meet the sailing moon
In awe i can't deny
Tonight at life's high noon.

The spate of regional to continent-scale fires, in Brazil, Siberia, California, southern Europe, Australia and elsewhere (Figs. 6.1, 6.2, 6.3, 6.4, 6.5, and 6.7), signifies portents of future calamities generated by temperature rises over tinder-dry regions of Earth where original forests, developed under

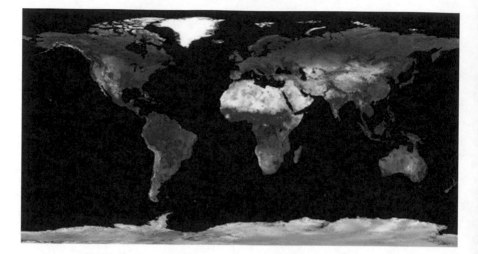

Fig. 6.1 The NASA earth data fire map accumulates the locations of fires detected by moderate-resolution imaging spectro-radiometer (MODIS) on board the Terra and Aqua satellites over a 10-day period. Each colored dot indicates a location where MODIS detected at least one fire during the compositing period. Color ranges from red where the fire count is low to yellow where number of fires is large

Mediterranean to sub-polar climate conditions, are overwhelmed by droughts and heat waves associated with the polar-ward migration of tropical and subtropical climate zones. As fires rage across tens of thousands square km the Amazon forest, dubbed the Planet's lungs, producing some 20% of the oxygen in the atmosphere. With some 72,000 fires in the Amazon this year, where fires on such a scale are uncommon, as well as through Siberia, Alaska, Greenland, southern Europe, California, Australia and elsewhere, increasing temperatures and droughts overwhelm original habitats, flora and fauna.

The Arctic Circle is suffering from an unprecedented number of wildfires in the latest sign of a climate crisis (Fig. 6.3). With some blazes the size of 100,000 football pitches, vast areas in Siberia, Alaska and Greenland are engulfed in flames. The World Meteorological Organization has said these fires emitted as much carbon dioxide in a month as the whole of Sweden does in a year. The world is literally on fire—so why is it business as usual for politicians?

"Climate change is making dry seasons longer and forests more flammable. Increased temperatures are also resulting in more frequent tropical forest fires in non-drought years. And climate change may also be driving the increasing frequency and intensity of climate anomalies, such as El Niño events that affect fire season intensity across Amazonia."

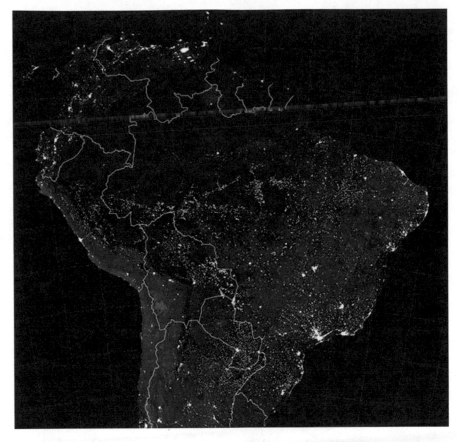

Fig. 6.2 Amazon fires. Note many fires follow clearing lines. NASA

According to Bowman et al. (2009), about 2–4 GtC are released annually from both natural and human triggered fires, intensified due to human activities, the rise in temperature and consequent droughts. Net emissions of anthropogenic greenhouse emissions increased by 35 percent from 1990 to 2010 EPA (2013). Approximately 42% of the added CO_2 stays in the atmosphere while the rest is sequestered by the ocean and vegetation (Global Carbon Project 2018). The total anthropogenic effective radiative forcing (ERF) over the Industrial Era is 2.3 Wm^{-2} (1.1–3.3 Wm^{-2}) (IPCC AR5). Combined with other greenhouse gases, this has led to an increase in the radiative forcing by ~+3.2 W m^{-2} translated to ~2.4 °C (Hansen et al. 2011).

Global warming and its disastrous consequences are now truly in Australia. At the moment a change in the weather has given parts of the country a respite from the raging fires, some of which are still burning or smoldering, waiting for another warm spell to flare up. The danger zones include

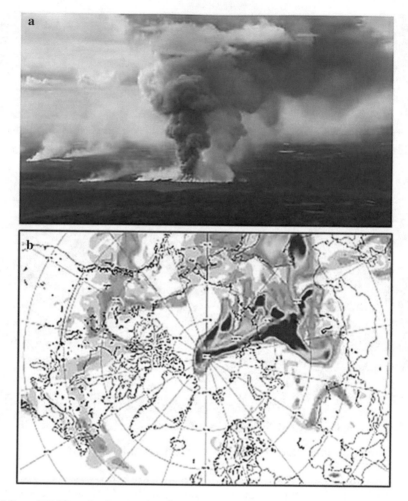

Fig. 6.3 **a** Wildfires in the Arctic often burn far away from population centers, but their impacts are felt around the globe. From field and laboratory work to airborne campaigns and satellites, NASA; **b** Thermal effects of aerosol form biomass burning in Siberia and the Arctic. NASA

the Australian Capital Territory, from where these lines are written. To date 186,000 square kilometers were burnt, including native forests, native animals, homesteads and towns, and 24 people died. The firestorms betray harbingers of a planetary future, or a lack of such, under ever rising temperatures and extreme weather events inherent in fossil fuel driven global warming.

In Australia mean temperatures have risen by 1.5 °C between 1910 and 2019 (Fig. 6.4a), as a combination of global warming and the ENSO conditions, as reported by the Bureau of Meteorology.

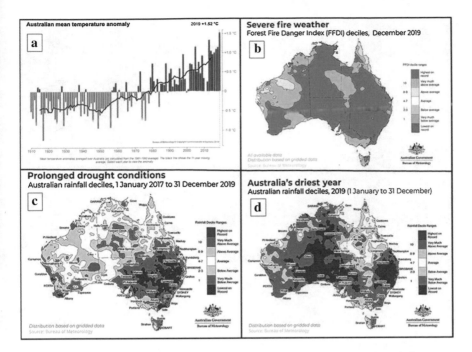

Fig. 6.4 **a** Australian temperatures; **b** severe fire weather; **c** drought; **d** driest year. Australian Bureau of Meteorology

Fig. 6.5 Australian fires 2019–2020. NASA

Fig. 6.6 Tasmanian fire

The Indian Ocean Dipole (IOD) has returned to neutral after one of the strongest positive IOD events to impact Australia in recent history … the IOD's legacy of widespread warm and dry conditions during the second half of 2019 primed the Australian landscape for bushfire weather and heatwaves this summer. In the Pacific Ocean, although indicators of the El Niño–Southern Oscillation (ENSO) are neutral, the tropical ocean near and to the west of the Date Line remains warmer than average, potentially drawing some moisture away from Australia.

The prolonged drought (Fig. 6.4c, d), low fuel moisture, high temperatures and warm winds emanating from the inland have rendered large parts of the Australian continent tinder dry, creating severe fire weather (Figs. 6.5, 6.6) subject to ignition by lightning and human factors. Fires on a large scale create their own weather (Fig. 6.6). Observations of major conflagrations, including the 2003 Canberra fires, indicate fires can form atmospheric plumes which can migrate and as hot plumes radiating toward the ground.

The underlying factor for rising temperatures and increasingly severe droughts in Australia is the polar-ward shift in climate zones as the Earth warms, where dry hot subtropical zones encroach into temperate zones, as is also the case in South Africa and the Sahara. **Smoke signals** emanating

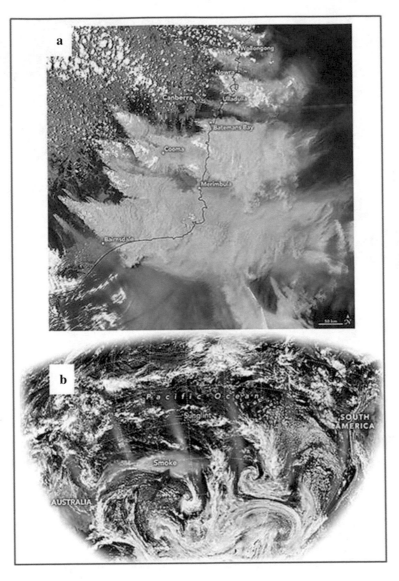

Fig. 6.7 **a** Smoke emanating from the southeastern Australian fires; **b** smoke from the pyro-cumulonimbus clouds of the Australian fires drifting across the Pacific Ocean. The fire clouds have lofted smoke to unusual heights in the atmosphere. The CALIPSO satellite observed smoke soaring between 15 and 19 km on January 6, 2020—high enough to reach the stratosphere. NASA

from the Australian fires are now circling around the globe (Figs. 6.7 and 6.8) signaling a warning of the future state of Earth should *Homo sapiens*, so called, not wake up to the consequences of its actions.

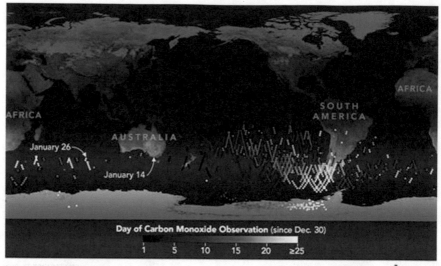

January 26, 2020   ⬇ JPEG

July 2019 - January 2020

Fig. 6.8 Australia is burning. Carbon monoxide emanations from the Australian bushfires, in billions of grams

7

The Gathering Storm

No one
Was there to hear
The muffled roar of an earthquake
Nor anyone who froze with fear
Of rising cliffs, eclipsed deep lakes
And sparkling comet-lit horizons
Brighter than one thousand suns
That blinded no one's vision.

No one
Stood there in awe
Of an angry black coned volcano
Nor any pair of eyes that saw
Red streams eject from inferno
Plumes spewing out of earth
And yellow sulphur clouds
Choking no one's breath.

No one
Was numbed by thunder
As jet black storms gathered
Nor anyone was struck asunder
By lightning, when rocks shattered
Engulfed by gushing torrents
That drowned the smoldering ashes
Which no one was to lament.

A. Y. Glikson, *The Event Horizon: Homo Prometheus and the Climate Catastrophe*,
https://doi.org/10.1007/978-3-030-54734-9_7

In time
Once again an orange star rose
Above a sleeping archipelago
Sun rays breaking into blue depth ooze
Waves rippling sand's ebb and flow
Receding to submerged twilight worlds
Where budding algal mats
Declare life
On the young earth

As the atmospheric concentration of the well-mixed greenhouse gases rise (CO_2 >411.76 ppm; CH_4 >1870.5 ppb ; N_2O >333 ppb plus trace greenhouse gases) land temperatures soar (NASA global sea-land mean of 1.05 °C since 1880). According to Berkeley Earth (2018) global land temperatures have increased by 1.5 °C over the past 250 years and mean Arctic temperatures have risen by 2.5–3.0 °C between 1900 and 2017. According to NASA:

1. Extreme heatwaves will become widespread at 1.5 °C warming. Most land regions will see more hot days, especially in the tropics.
2. At 1.5 °C about 14% of Earth's population will be exposed to severe heatwaves at least once every five years, while at 2.0 °C warming that number jumps to 37%
3. Risks from forest fires, extreme weather events and invasive species are higher at 2.0 °C warming than at 1.5 °C warming
4. Ocean warming, acidification and more intense storms will cause coral reefs to decline by 70–90% at 1.5 °C degrees Celsius warming, becoming all but non-existent at 2.0 °C warming.

Bar the transient masking effects of sulphur aerosols, which according to estimates by Hansen et al. (2011) induce sun shields of more than 1.0 °C of warming, global temperatures have already reached near 2.0 °C.[1] As aerosols are not homogeneously distributed, in some parts of the world temperatures have already soared to such levels. Thus the degree to which aerosols cool the earth, which depends on aerosol particle size range, has been grossly underestimated.

The rate of global warming, at ~2–3 ppm year and ~1.5 °C in about one century, faster by an order of magnitude then geological climate catastrophe

---

[1] By analogy to the requirement for a patient's body temperature to be measured before and not after aspirin has been taken.

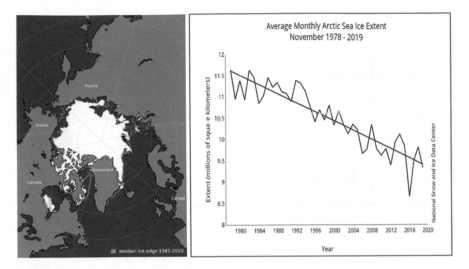

Fig. 7.1 The reduction in Arctic sea ice between 1978 and 2019, NASA

such as the PETM and the KT (Kring 2000, 2002) impact, has taken scientists by surprise, requiring a change from the term *climate change* to *climate calamity*.

The prevailing political and economic focus of international conferences and elsewhere is mainly on (1) Limits on, or a decrease of carbon emissions from power generation, industry, agriculture, transport and other sources; (2) limits on the current rise in global temperatures to +1.5 °C and a maximum of +2.0 °C above mean pre-industrial (pre-1750) temperatures.

However, at the present the concentration of greenhouse gases of near ~500 ppm CO_2-equivalent is activating amplifying feedbacks of greenhouse gases from land, oceans and melting ice sheets.

1. An increase in evaporation due to warming of land and oceans leads to further warming due to the greenhouse effect of water vapor but also to increased cloudiness which retards warming. The water vapor factor, significant in the tropics, is less significant in the dry subtropical zones and relatively minor in the Polar Regions.
2. The melting of ice sheets, reducing reflective (high-albedo) ice and snow surfaces and concomitantly opening of ocean surfaces (heat absorbing low-albedo surfaces) is generating a powerful positive (warming) feedback. Hudson (2011) estimates the rise in warming due to total removal of Arctic summer sea ice as approximately +1 °C (Fig. 7.1), along with increasing rates of melt of the Greenland ice sheet (Fig. 7.2).

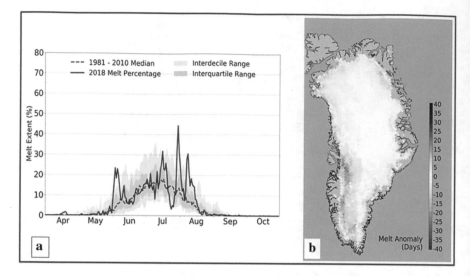

Fig. 7.2 a Surface melt area as a percentage of the ice sheet area during 2018 (solid red), in addition to the 1981–2010 average (dashed blue) and inter-decile and interquartile ranges (shaded); **b** melt anomaly (in number of melting days) with respect to the 1981–2010 period during the summer of 2018 estimated from space-borne passive microwave observations. NOAA Arctic program

3. The release of methane from melting permafrost and bubbling of methane hydrates from the oceans has already raised atmospheric methane levels from about 800 to 1863 ppb which, given the radiative forcing of methane of $X_{25}CO_2$, is significant (Fig. 7.3).

4. As the oceans warm they become less capable of taking up carbon dioxide. As a result, more of our carbon pollution will stay in the atmosphere, exacerbating global warming.

5. As tropical and subtropical climate zones overtake temperate Mediterranean-type climate zones, desiccated and burnt vegetation releases copious amounts of carbon dioxide to the atmosphere. For example the current bushfires in Eastern Australia (Figs. 6.5–6.7) have already emitted 250 million tonnes of CO_2, almost half of the country's annual emissions in 2018.

   With rising global temperatures and further encroachment of subtropical climate zones (Fig. 5.5), reductions in emissions may be insufficient to stem global warming, unless accompanied by sequestration of greenhouse gases, recommended as below 350 ppm CO_2. Hansen et al. (2008) see significant potential in the biochar method, utilizing pyrolysis of residues of crops, forestry and animal waste. Biochar helps soil retain nutrients and fertilizers,

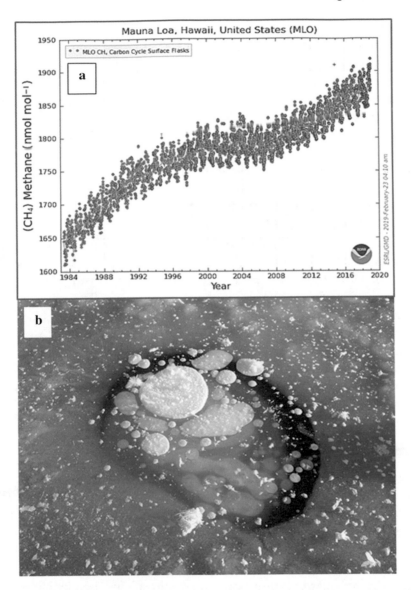

Fig. 7.3 a The rise in methane between 1984 and 2020 at Mauna Loa. NASA;
b Frozen Methane Bubbles. When ice-rich permafrost thaws, former tundra and
forest turns into a thermos-karst lake as the ground subsides. The carbon stored
in the formerly frozen ground is consumed by the microbial community, who release
methane gas. When lake ice forms in the winter, methane gas bubbles are trapped
in the ice. Location: Alaska. Credit: Miriam Jones, USGS

reducing release of greenhouse gases such as N_2O. Replacing slash-and-burn agriculture with a slash-and-char method and using agricultural and forestry wastes for biochar production could provide a CO_2 drawdown of ~8 ppm or more in half a century.

No one knows how to impose 1.5 or 2.0 degrees Celsius limits on the mean global temperature, unless drawdown/carbon sequestration of atmospheric CO_2 can be attempted (Table 7.1; Fig. 7.4), nor are drawdown methods normally discussed in most political or economic forums. However, no one knows how to impose these limits, if drawdown/sequestration of atmospheric CO_2 is not attempted. According to Drum (2019) *"Meeting the climate goals of the Paris Agreement is going to be nearly impossible without removing carbon dioxide from the atmosphere"* (Fig. 7.4).

Stabilization and cooling of the climate would include two principal approaches (Table 7.1): (A) solar shielding, and (B) CO_2 sequestration (Fig. 7.4). However, solar shielding by injected aerosols or water vapor is bound to be transient, requiring constant replenishment. The question is how effective could these methods be in reducing CO_2 levels on a global scale, at the very least to balance emissions, currently 36.8 billion tons CO_2 per year. It is hard to see an alternative way of cooling the atmosphere and oceans than a combination of several of these methods (Table 7.1). Budgets on a scale of military spending ($1.7 trillion in 2017) would be required in an attempt to slow down the current trend across climate tipping points. The choice humanity is facing is whether to spend resources of this scale on wars or on defense from the climate calamity.

The widest chasm has developed between what climate science indicates and climate policies and negotiations controlled by governments, politicians, economists and journalists—none of whom fully comprehends, or is telling the whole truth about, the full consequences of the calamitous trend in the atmosphere–ocean–land system. Future climate projections outlined by climate science have been largely put aside mainly since it is economically and politically "inconvenient" or too frightening.

Over the last quarter century carbon emissions have risen by almost 63% (1990–22.6 $MtCO_2$/year; 2005–30 $MtCO_2$/year; 2017–37 $MtCO_2$/year). As a consequence of global carbon emissions, by 2018 mean global temperatures reached +0.98 °C above pre-industrial conditions and further by more than +0.5 °C over the continents, for example reaching +2.2 °C in Mongolia. The transient cooling effect of human-emitted aerosols potentially ameliorates between 0.5 and 1.0 °C, as has been manifested for example when flights' contrails were discontinued on 9/11.

Table 7.1 Solar shielding and atmospheric CO_2 sequestration methods

Method	Supposed advantages	Problems
SO_2 injections	Relatively cheap and rapid application	Short atmospheric residence time; ocean acidification; retardation of precipitation and of monsoons
Space satellite-mounted sunshades/mirrors	Rapid application. No direct effect on ocean chemistry	Longer space residence time. Does not mitigate ocean acidification by CO_2 emissions
Streaming of air through basalt and serpentine (Fig. 4)	CO_2 capture by Ca and Mg carbonates	In operation on a limited scale in Iceland. *Significant potential*
Soil carbon burial/biochar	Effective means of controlling the carbon cycle (plants + soil exchange more than 100 GtC/year with the atmosphere)	Requires a collaborative international effort by millions of farmers. *Significant potential*
CO_2 capture by seaweeds	An effective method applied in South Korea	Decay of seaweeds releases CO_2 to ocean water. *Significant potential*
Ocean iron filing fertilization enhancing phytoplankton	CO_2 sequestration	Phytoplankton residues would release CO_2 back to the ocean water and atmosphere
Ocean pipe system for vertical circulation of cold water to enhance CO_2 sequestration	CO_2 sequestration	Further warming would render such measure transient
"Sodium trees"—pipe systems of liquid NaOH sequestering CO_2 to sodium carbonate Na_2CO_3, followed by separation and burial of CO_2	CO_2 sequestration, estimated by Hansen et al. (2008) at a cost of ~$200/ton CO_2 where the cost of removing 50 ppm of CO_2 is ~$20 trillion	Unproven efficiency; need for CO_2 burial; $trillions expense, though no more than the *military expenses since WWII*

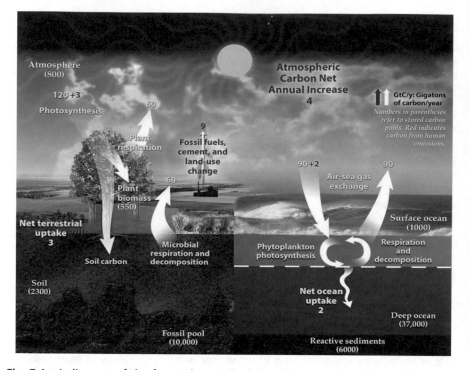

Fig. 7.4 A diagram of the fast carbon cycle shows the movement of carbon between land, atmosphere, soil and oceans in billions of tons of carbon per year. Yellow numbers are natural fluxes; red are human contributions in billions of tons of carbon per year. White numbers indicate stored carbon

The melting of the polar ice sheets, where warming takes place at twice the rate as the rest of the Earth, heralds a further fundamental global climate shift, rendering even larger parts of the land subject to extreme weather events such as those already affecting island chains and coast regions, costing the lives of tens of millions.

Life on Earth is controlled by the presence of water, insolation, the composition of the atmosphere and oceans, asteroid impacts and large volcanic eruptions. Cyclic and abrupt changes in these factors have affected the climate over billions of years. Sharp rises in atmospheric greenhouse gas concentrations, such as about 66 and 56 million years ago, have led to major crises in the biosphere. The current rise in atmospheric greenhouse gas concentration, combining the effects of CO_2, methane and nitric oxide, is now tracking toward 500 ppm CO_2 equivalent (Fig. 5.2), the stability threshold of the large ice sheets. The consequent rise in mean global temperature by 3 to 4 degrees Celsius over a period as short as a century or two represents the highest recorded in geological history.

8

The Critical Century

We play Russian roulette with the climate and no one knows what is in the active chamber of the gun (Wallace Broecker 1931–2019).

Two years after moving the minute hand of the Doomsday Clock to within two minutes of midnight—a warning for all humanity—the science and security board of the Bulletin of the Atomic Scientists moved the minute hand to 100 seconds before midnight (Welna 2020). Never since the 1947 Cold War has the clock been set so close to the putative doomsday annihilation. All that while the scale and pace of global warming are astounding climate scientists (Oreskes et al. 2019). Despite its foundation in the basic laws of physics—the black body radiation laws of Planck, Kirchhoff' and Stefan Boltzmann, and empirical climate observations around the world by climate research organizations—NOAA, NASA, NSIDC, IPCC, World Meteorological Organization, Hadley-Met, Tindale, Potsdam, BOM, CSIRO and others, the anthropogenic origin of climate change remains a subject of denial and untruths. As the world is heating fire-fighting defenses are superseded by orders of magnitude by the resources allocated to the military, aimed at yet another catastrophe.

Throughout history messengers of bad news have been rebuked or worse. The Cassandra syndrome (Fig. 8.1) is alive and well, climate scientists finding themselves in a predicament similar to that of medical doctors committed to convey a grave diagnosis but finding it difficult to communicate the warning, even among friends and family. Parliaments, preoccupied with economics,

Fig. 8.1 Ajax raping Cassandra from the Palladium. Attic black-figure Kylix, ca. 550 BC. *Public Domain*

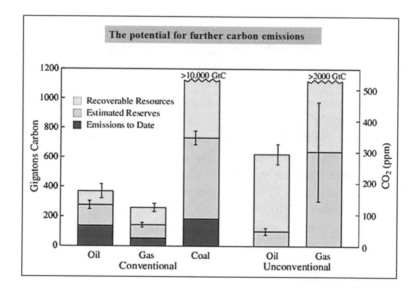

Fig. 8.2 Estimates of fossil fuel resources and equivalent CO_2 levels, including (1) emissions to date (~2.12 GtC = 1 ppm CO_2); (2) estimated hydrocarbon reserves, and (3) recoverable hydrocarbon resources (Hansen et al. 2013c). Courtesy James Hansen

legal issues, domestic and international conflicts, hardly regard the future of nature and civilization as a priority, nor is the media willing to communicate the dire climate projections.

Humans are earthling, from and of the Earth, physically adapted to its climates, gravity, radiation, electro-magnetic field and the chemistry of the atmosphere. Had there ever been a single critical dictum science has conveyed to *H. sapiens* it is that altering the composition of the Earth's atmosphere can bear fatal consequences. Yet as new coal reserves are opened (Fig. 8.2), new

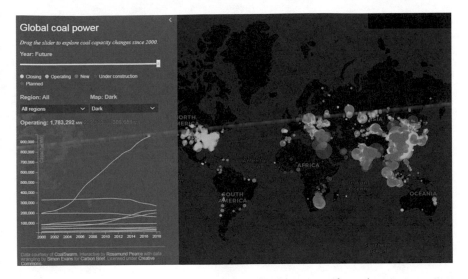

Fig. 8.3 Global coal combustion plants: Distribution of coal power generators: yellow—operating (1.783, 292 MW); orange—new; red—under construction (232,133 MW); Purple—planned (306,651 MW); *Creative commons*

power plants are under construction (Figs. 8.3 and 8.4) and carbon emissions grow at an accelerated rate, the race to destroy species and civilization is heading toward its conclusion.

The message communicated by scientists since the 1980s has become the subject of obfuscation by vested interests, aided by much of the media. Many understand but feel powerless, voting for parties that, under false promises take little or no effective measures at reduction of hydrocarbon mining and carbon emissions. False expressions such as *"a one in 100 years event"* are common, nor do the terms *"climate change"* or *"global warming"* convey the extreme rate of climate disruption. Much of the debate overtaken by politicians and economists, mostly to the exclusion of climate scientists, conveys little understanding of physical processes in the atmosphere–ocean system. *Infotainment* programs where attractive celebrities promote space travel distract people from the climate emergency.

According to Andrews (2019) *"The planet is warming, the oceans are acidifying, the Amazon is burning down, and plastic is snowing on the Arctic. Humanity's environmental devastation is so severe, experts say, that a global-scale ecological catastrophe is already underway. Even those holding sunnier views would be hard-pressed to deny that our global footprint is presently less a light touch and more a boot stamping on Earth's face. Against this dark background, one might ask if spending lavish sums to send humans to other worlds is a*

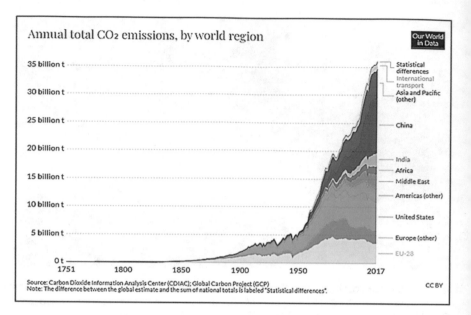

Fig. 8.4 Growth of fossil fuel CO_2 emissions, World Data 1751–2017. Since 1751 the world has emitted over 1.5 trillion tonnes of CO_2. To reach our climate goal of limiting average temperature rise to 2 °C, the world needs to urgently reduce emissions. Creative Commons

foolhardy distraction—or a cynical hedge against life's downward *spiral on this one"*.

The UNHCR estimated that since 2008 some 22.5 million people have been displaced by climate or weather-related events. According to researchers from the World Health Organization (WHO) and the University of Wisconsin global warming is already responsible for some 150,000 deaths each year around the world. In 2014, WHO estimated climate change would lead to about 250,000 additional deaths each year between 2030 and 2050, from factors such as malnutrition, heat stress and malaria.

The climate crisis constitutes the biggest assault on nature since 66 million years ago and, except for the nuclear threat, the greatest existential danger humanity has ever faced. From 1870 to 2014, cumulative carbon emissions totaled about 545 GtC. Approximately 230 GtC (or 42%) of the emissions were retained in the atmosphere, about 155 GtC (or 28%) absorbed by the oceans and approximately 160 GtC (or 29%) by land. Future burning of the known fossil fuel reserve (Fig. 8.2) would transfer an additional >3000 GtC to the atmosphere, threatening to render large parts of the Earth uninhabitable.

Apart from lip service and non-binding agreements, governments are presiding over the demise of much of the global biosphere and of civilization. As the planet keeps warming and the powers spend trillions of dollars on so-called "defense" and murderous wars, people and nature suffer and die.

While lethal gases spew into the atmosphere toxic untruths inundate the population, compliments of lie factories and the $trillions advertising industry. The Advertising and Marketing industry saw huge growth in 2017, and now comprises a market value of $1.2 Trillion, pushing commercial stunts that bear no relation to the product, political ruses of no relation to the message, ecstatic faces rejoicing in new soap powder, jumping with joy at new model cars, consoled by funeral insurance policies. Young generations brought up in front of electronically propagated lies are growing unable to separate truths from lies.

But a new generation is emerging, inspired by the life-giving message of a 16 years-old school girl—Greta Thunberg (Fig. 1.1).

Every week the internet reports newly discovered and exploited coal, oil and gas fields and at the same time reports increasingly intense cyclones, floods and fires. Whereas any single extreme weather event may or may not be related to global warming, the tripling of the incidence of extreme weather events (Fig. 5.5) heralds the warming of the planetary system, taking the world away from conditions that allowed humans to flourish since the Neolithic age. Intense cyclones and flood events are hitting the Caribbean islands, southeastern Texas, Florida, Mississippi, southwest Pacific islands, the Philippines, Kerala, Mozambique, Zimbabwe, Japan, Queensland, and elsewhere. While promises of clean renewable energy abound mining and export of coal, oil and gas multiply (Fig. 8.3)—it all ends up in the same atmosphere (Fig. 8.4).

It is estimated that, to date, climate change has been linked to 400,000 deaths worldwide each year, projecting deaths to increase to close to 600,000 per year by 2030 (Leber 2015). This includes, for example, 1833 people in New Orleans, possibly up to 5000 in Puerto Rico, 6329 by typhoon Haiyan in the Philippines. A quantitative risk assessment by the World Health Organization (WHO) suggests additional climate change-related deaths for the year 2030 would total 241,000 people, 38,000 due to heat exposure in elderly people, 48,000 due to diarrhea, 60,000 due to malaria, and 95,000 due to childhood under-nutrition.

A changing climate not only affects agriculture but also leads to greater food spoilage from heat, including diarrheal illnesses and hunger that caused around 310,000 deaths in 2010. Heat and cold illnesses, malarial and vector-borne diseases, meningitis and environmental disasters account for the rest of

Fig. 8.5 Industrial smoke stacks whose emissions form a hurricane eyewall. Poster for the film. *An Inconvenient Truth*

the almost 700,000 deaths attributable to these direct climate impacts. Pollution, indoor smoke, and occupational hazards related to the carbon economy cause 5 million deaths through ailments like heart disease, skin and lung cancer.

The first casualty of war is the truth. Throughout history, conspiracies, false flag attacks and denial led to catastrophe. On a global scale, by analogy with the tobacco denial syndrome, the fossil fuel industry has been paying denialists and pseudo-scientists for propagating untruths about the origins and consequences of global warming (Fig. 8.5). Irrefutable evidence denied by pro-coal lobbies and compliant politicians include:

- Denial of basic laws of physics, i.e. the blackbody radiation laws of Plank, Stefan-Boltzmann and Kirchhoff.
- Denial of direct observations and measurements in nature, in particular the sharp rises of temperatures, ice melt rates, sea level rise and extreme weather events.
- Denial of the global warming origin of extreme weather events, i.e. the closely monitored rise in storms, hurricanes, fires and droughts in several parts of the world.
- Denial of the bulk of the peer-reviewed literature summed up in the IPCC reports.
- Denial of conclusions of the world's premier climate research organizations (NASA, NOAA, NSIDC (National Snow and Ice Data Centre), Hadley-Met, Tindale, Potsdam, World Meteorological Organization (WMO), CSIRO, BOM and other organizations).

The Faustian Bargain[1] (Fig. 8.6) is alive and well, where the world's wealthiest enterprises, the fossil fuel industry, billionaires, captains of industry and their political and media mouthpieces are devastating the Earth atmosphere and leading toward nuclear wars, while the rest of the world is obsessed by bread and circuses.

According to Hansen et al. (2013):

Humanity is doubling down on its Faustian climate bargain by pumping up fossil fuel particulate and nitrogen pollution. The more the Faustian debt grows, the more unmanageable the eventual consequences will be. Yet there are plans to build more than 1000 coal-fired power plants and plans to develop some of the dirtiest oil sources on the planet. These plans should be vigorously resisted. We are already in a deep hole – it is time to stop digging.

According to Julian Cribb "*current trends include blind support for fossil fuels, overt denialism or a reluctance to act on climate, prejudice … a tendency to favor ecological rapine and increased pollution*". While caravans of desperate people flee gangster-ridden Central American republics, to be blocked by razor-wired walls, starving Africans suffocate in air-lock trucks across the English Channel and asylum seekers commit suicide in Pacific concentration camps.

---

[1]Faustian bargain, a pact whereby a person trades something of supreme moral or spiritual importance, such as personal values or the soul, for some worldly or material benefit.

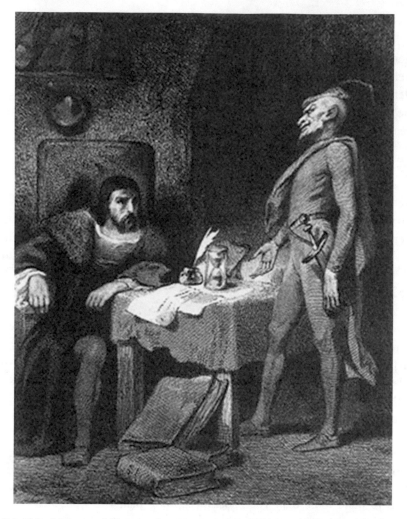

Fig. 8.6 Mephisto and Doctor Faustus contemplating a bargain with the devil, engraving by Tony Johannot

Hoodwinked by half-truths propagated by privately-controlled conscience-free politicians and the media, the inhabitants of suburbia international are lured by Eurovision circuses, royal weddings, cricket ball tampering and promises of space tourism and other distractions. Around the world, according to those who portray "black is white" and believe in "Good vs Evil", "Good" constitutes brute force and dominance they see as admirable and "Evil" constitutes cooperation and compassion which they see as weakness. When the term "*future*" is expressed in the media and parliaments it betrays no idea as to what kind of future holds under +2 or +4 degrees

Fig. 8.7 The Creation of Adam (Italian: Creazione di Adamo) is a fresco painting by Italian artist Michelangelo, which forms part of the Sistine Chapel's ceiling, painted c. 1508–1512

Celsius above pre-industrial temperatures, rendering large parts of the planet uninhabitable.

While temperatures keep rising according to mindless scorched Earth policies, the human sense of reality becomes blurred by electronic system, symbolized by the computer worship (*apes with apps*) succeeding the divine touch (Fig. 8.7), the internet, *TV or not to Be*, science fictions, virtual realities, mass circuses, world number#1 celebs, sport legends, space travel, fake news, glitzy shows, canned laughter, gratuitous hype, superlatives and decrepit brutality, while factories of death spewing deadly ash (8.8, 8.9) are replaced by carbon dioxide death factories (Figs. 8.3 and 8.4), aided by $trillions governments subsidies, igniting global-scale fires destroy much of the land (Figs. 6.1 and 6.2).

While the seas are encroaching toward coastal plains, delta and low river valleys, where the bulk of the world's population live and grow food, studies of the last interglacial (Eemian ~130 to ~118 kyear) indicate the world is already committed to 6–9 m sea level rise. No one knows whether the rise in global temperature has already reached a point of no return, whether emission levels can be halted and negative emissions sequestered on a scale that can slow down, arrest or even reverse global heating. Would the global Empires and their satellites be willing to address the growing calamity? Will the elites keep diverting $trillion-scale resources from the military into the defense of life on Earth?

Fig. 8.8 The Auschwitz gas chambers

With estimated carbon reserves in excess of 20,000 GtC, over 20 times the CO_2 content of the atmosphere, further emissions can take the atmosphere above 1000 ppm CO_2, namely to Early Eocene (~54–48 Ma) conditions when CO_2 levels are inferred to have been between 1,000 and 2,000 parts per million and large parts of the continents were inundated by the oceans. Where WWII sacrificed millions in gas chambers (Fig. 8.8), global warming threatens to destroy billions (Fig. 8.9) on the strength of an *"economic"* Faustian Bargain.

The peril posed by the two largest industries of death, i.e. fossil fuels and the military, are intertwined. In his book The Doomsday Machine Daniel Ellsberg (2017) documents the most dangerous arms buildup in the history of civilization, threatening the very survival of humanity. No other insider has reported so candidly on the long-classified history of the nuclear arms race, intensifying at present. Although in principle the advent of nuclear weapons was supposed to render wars obsolete, at present no negotiations are held between the big powers on limiting nuclear weapons or removing them from on-alert state. In the absence of ideological differences between west and east, both dominated by billionaires, nor does any rationale exist for nuclear weapons, controlled by computers, which appear to have assumed an independent life of their own. With time possibilities become probabilities become certainties, pushing the world toward a suicide pact by accident or design.

Fig. 8.9 Global carbon dioxide poisoning. Cholla power plant near Joseph City, Arizona

From the Romans to the third Reich, the barbarism of empires surpasses that of small marauding tribes. In the name of their gods and nation, where "might is right", empires never cease to kill poor people in their small fields. It is among the wretched of the Earth that true charity is found, where empathy is learnt through suffering. The battle between life-enhancing and death-inducing agents in nature, symbolized by the Brahma-Vishnu-Shiva trinity, never ceases. In our time a blood-stained civilization buzzing with witless twits and faceless books is once again overtaken by blind extremism, this time leading to a radioactive hothouse Earth. Survivors may suffer, as they have during the extreme climate upheavals of the glacial-interglacial cycles, enduring through the most adverse conditions.

Leaders, so called, are not telling the truth, either because they do not understand or perhaps because they do. It is not clear what turns aspiring young frontrunners, once they achieve power, to become turncoats betraying their original calling. Is it ambition, competitive urge, clan loyalty on a sinking ship? In the lack of leadership power resides in the *Fourth Estate*, dominated by ratings and infotainment, acting as prosecutor, judge and jury on behalf of commercial interests and self-appointed ideologues, thriving on conflict and vilifying perceived enemies. News of climate events, given lower priority than sports and entertainment, are rarely accompanied with explanation of the full consequences of the toxic effects of carbon gases on the biosphere.

In long term, since the Neolithic, human civilization has perpetrated, unaware or aware, a war against the forests, other species and nature itself. Should anyone be there to record the history of the twentieth and twenty-first centuries they would observe that, while temperatures and sea levels were rising, people's sense of reality has been blurred by the multi-$billion commercial and political advertising industry and by electronic systems. Since WWII the internet, the anti-social media (Wilson 2020), smart phones, Face-less books (Naughton 2018), Twits, virtual realities, public circuses, fake news, celebrity icons, superlatives, hype and drugs have been swamping common sense, inducing an atmosphere of unreality. An endless stream of homicide shows, ruled by the eternal couple of sex and war, dominates a culture of death. Truth, in the eye of the beholder, belongs to those individuals who succeed in keeping a degree of sanity, and to the suffering multitudes in distant parts of the Earth (Fig. 8.10) targeted by missiles and killer drones (Piper 2019), and not to those who watch their anguish on television screens much as the Romans entertained themselves by gladiator games.

According to Mary Debrett: *"We are now in the middle of a perfect storm of mis-communication regarding climate change. Various factors have converged to confound rational public conversation. Public opinion polling indicates that although there is widespread acceptance of climate change resulting from human activities, the public's preparedness to pay for action to mitigate climate change is actually declining—even as climate scientists warn of the increasing urgency for acting"*.

International climate conferences (COP-25, Madrid 2019) and advisory councils are focused on (A) limits on, or a decrease of, carbon emissions from, power generation, industry, agriculture, transport and other sources; (B) limits on the current rise of global temperatures to +1.5 °C and a maximum of +20 °C above mean pre-industrial (pre-1750) levels. Little account is taken of amplifying feedback loop induced by the rising temperatures. This includes further warming driven by the replacement of reflective ice and snow surfaces by infrared-absorbing open water, by methane release from permafrost and methane clathrates, by extensive fires, and by reduced CO_2 absorption by warming oceans, and by fires, droughts and reduced photosynthesis.

Fig. 8.10 The real world: **a** Kenya. IOM 2012—MKE0666; **b** Haiti. IOM 2014—MHT0593

Given the cumulative long atmospheric residence time of CO_2 (Solomon et al. 2009; Eby et al. 2009) and the short life span of aerosols, attempts at CO_2 drawdown are essential. Abrupt reductions in emissions may be insufficient to stem global warming, unless accompanied by sequestration of greenhouse gases from the atmosphere, required to below 350 ppm CO_2 (Hansen et al. 2008). This would require carbon sequestration by soils according to the biochar method (Cooperman 2016) (Fig. 8.11), involving pyrolysis of residues of crops, forestry and animal waste. Biochar helps soil retain nutrients and fertilizers, reducing release of greenhouse gases such as N_2O. According to Hansen et al. (2008) replacing slash-and-burn agriculture with a slash-and-char method and the use of agricultural and forestry wastes for biochar production could provide a CO_2 drawdown of ~8 ppm or more in half a century.

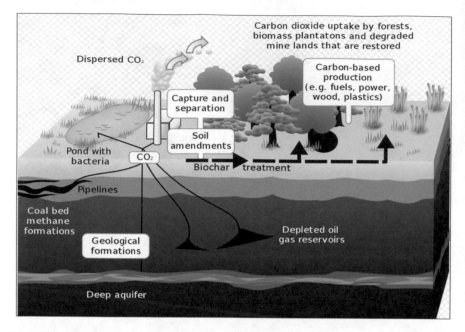

Fig. 8.11 Schematic presentation of the carbon cycle including biochar soil amendment. CO_2 combines with water to form carbonic acid (H_2CO_3) that subsequently loses hydrogen ions (H+) to form bicarbonate (HCO^{3-}) and carbonate (CO_3^{2-}) ions

In the three years following the Paris Agreement (2016–2018), taking little notice of international agreements, US, Canadian, European, Chinese and Japanese banks have provided US$1.9 trillion for the extraction and transfer to carbon from the Earth to the atmosphere, including coal mining, tar sands oil, Arctic oil and gas, deep water oil and gas drilling, coal seam gas fracking and liquefied natural gas (Figs. 8.3 and 8.4).

In medieval times poisoning of wells constituted a hanging offence. Nowadays, despite overwhelming evidence, saturation of the atmosphere with greenhouse gases and acidification of the oceans constitute the source of the highest profits. For over 40 years the powers that be, fully cognizant of stern warnings by scientists (Carrington 2019), have continued to allow and enhance the transfer of carbon from the Earth to the planetary atmosphere, leading to extreme weather events and sea level rise, with fatal consequences for the habitability of large parts of Earth (Wallace-Wells 2017).

It is not a coincidence that political movements that promote racism and war are oblivious to the dissemination of poisoned power, including carbon saturation of the atmosphere and dissemination of radioactive isotopes on land and water, commonly citing "economic" reasons. However, there can be no "economy" in a + 4 degrees Celsius world (New et al. 2011). A large group pays lip service to the reality of the climate calamity but then, once in power, rarely places limits on emissions.

9

An Orwellian Climate as Rome Burns

Time's a silver wing in the rain
A railway station bench
Will I ever see you again?
Will my thirst ever quench?
Time you caught up with me
Pull me back into the sea
Bury me under my tree
Branching to infinity
Future's recess to memories
Can't cease my headlong quest
Wind blows out candles, trees
Will grow where I will rest

The people have no voice since they have no information (Gore Vidal).

The response to the climate calamity, ranging from warning by scientists, to deep concern by those who love nature, to utter denial and abuse by vested interests and their agents, has become the litmus test identifying the roots and ultimate nature of human conscience.

In George Orwell's novel 1984 the concept of *Newspeak* constitutes an anti-thesis of evidence-based reality where *Newspeak* is defined as "*a controlled language of restricted grammar and limited vocabulary, meant to limit the freedom of thought—personal identity, self-expression, and free will—that*

A. Y. Glikson, *The Event Horizon: Homo Prometheus and the Climate Catastrophe*,
https://doi.org/10.1007/978-3-030-54734-9_9

Fig. 9.1 George Orwell. *Who controls the past controls the future. Who controls the present controls the past*

Fig. 9.2 The arithmetic of an alternative logic: 2 + 2 plus the enthusiasm of the workers = 5". Soviet propaganda poster by Yakov Guminer, 1931

threatens the ideology of the regime of Big Brother and the Party, who have criminalized such concepts into thought-crime". Thought threatens the ideology of the regime of Big Brother and the Party, who have criminalized such concepts into *thought crime* (Figs. 9.1 and 9.2). The party's slogan *Who controls the past controls the future, who controls the present controls the past,* is repeated in our time in an attempt to wipe memories, including Holocaust denial and the Hiroshima-Nagasaki atom bombs, as well as the memory of a times when core values proclaimed *all men are created equal.*

History can be repeated in cycles. Wars are followed by periods of rebuilding and every few decades a new generation emerges which forget the lessons from the last collective bloodshed. According to *Newspeak* to condition people's minds for the next atrocity history must be eliminated and the language and meaning of words changed, altering people's way of thinking, where "*2 + 2 = 5 if the party says so*" (Fig. 9.2).

For all the horrors documented in Orwell's novel "1984" the destruction of nature is not the focus, which assigns the climate disruption and the nuclear menace of modern times to a special category of horror, as indicated by the clock of the atomic scientists at 100 s to mid-night.

20–21st Century *Newspeak* terms include:
Truth = When a lie is told enough times.
Democracy = When every dollar has an equal vote.
Defense = Preparations for the next war.
Security = When a computer chip failure can cost the Earth.
National security = A rationale for violation of human rights.
Victory = The body bag count.
The share market = When a wing flutter triggers economic collapse.
Conversation = A pre-determined one-sided decision.
Right vs wrong = As decided by the powers that be.
Globalization = Domination by international corporations.
Fake News = Disinformation manufactured by vested interests.
Economic rationalism = Where everything has a price, including life.
Sustainability = A cover-up term for business-as-usual.
The natural environment = Secondary to the business-environment.
GNP Growth = The psychology of a cancer cell.
Morality = Might is right.
World Number One = First in the popularity rating.

Inherent in fascism are extreme nationalism, totalitarianism, racism, militarism and a push for war, all of which are on the increase, as reflected by the mainstream media. Not uncommonly the *Fourth Estate* excels itself in derogatory comments aiming at perceived adversaries, personal smears and unverified false flags. In her book Fascism: A Warning Madeleine Albright states the Fascism of a century ago was not atypical: "*In hindsight, it is tempting to dismiss every Fascist of this era as a thoroughly bad guy or a lunatic, but that is too easy, also dangerous,*" she writes. "*Fascism is not an exception to humanity, but part of it ... Anti-democratic leaders are winning democratic elections ... and some of the world's savviest politicians are moving closer to tyranny with each passing year"*.

Only rarely have the victims of the industrial-scale wars rebelled against their masters. During 1914 and 1918 the Great War killed or wounded more than 25 million people peculiarly horribly. There were odd moments of hope in the trenches, when for a few brief hours men from both sides on the Western Front laid down their arms, emerged from their trenches, and shared food, carols, games and comradeship, in defiance of instructions (Fig. 9.3a).

For the generation that survived World War II, growing under the shadow of the atomic cloud, hopes for re-emergence of humanism, enlightenment, justice and peace did not last. In so far as it has been assumed the growing extreme weather events will cause people to realize the consequences of carbon emissions, obfuscation by the mainstream media has proved effective.

According to Noam Chomsky *"Humanity faces two imminent existential threats: environmental catastrophe and nuclear war. These were virtually ignored in the campaign rhetoric and general coverage. There was plenty of criticism of the Trump administration, but scarcely a word about by far the most ominous positions the administration has taken: increasing the already dire threat of nuclear war, and racing to destroy the physical environment that organized human society needs in order to survive"*.

In the wake of Benito Mussolini's adoption of *fasces*[1] the term fascism (Fig. 9.4) has acquired a universal significance, encompassing a range of malignant ideologies—characterized by hate of minorities, immigrant, colored-skin people, Jews, gypsies, socialists, intellectuals, scientists and the disabled, and the promotion of violence, misogyny, defense of the wealthy and promotion of poisoned power—coal and nuclear. At the heart of fascism is the final stage of the Brahma-Vishnu-Shiva cycle. Fascists may not worry about the genocidal consequences of their ideology, perhaps assuming they and their rich benefactors may survive the consequences.

A seamless transition exists between so-called conservative and pre-fascist parties, including enthusiasm for nuclear weapons and objection to a reduction of carbon emissions, relegating their philosophies to a culture of death. Fascism did not invent the climate catastrophe, but when faced with it is displaying a lack of concern regarding the fate of populations and of nature. Deep schisms pertain between those trying to defend nature and life and those who advocate conflict and war. Not all those who deny the climate evidence are of the extreme right, but such attitudes betray greater concern

---

[1] A Roman symbol consisting of a bundle of birch rods combined with an axe.

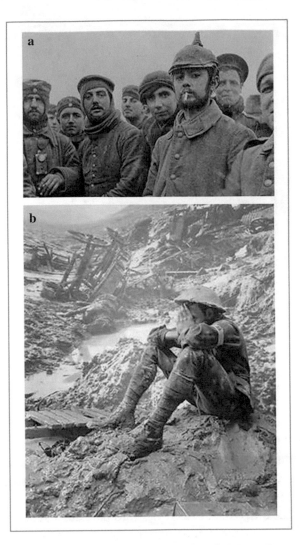

Fig. 9.3 Industrial war: **a** British troops during the Christmas Truce with Saxons of the 104th and 106th Regiments of the Imperial German Army; approximately 100,000 German and British troops were involved in unofficial ceasefires along the Western Front. On Christmas Eve, German troops decorated the areas around their trenches, continuing their celebration by singing Christmas carols. Soon after, British troops began singing carols of their own, followed by both sides exchanging goods such as tobacco and alcohol in the middle of the battling trenches at no man's land. In many of the sectors, celebrations lasted through Christmas evening all the way until New Year 1915; **b** A first-aid soldier in a bomb crater on the western front

Fig. 9.4 Upper part: Eagle with fasces used on the war flag of the fascist *Republica Sociale Italiana;* lower part: Gold coin from Dacia, minted by Coson, depicting a consul and two lictors (officials)

for material economic growth than for the natural environment. But fascism is far from exclusive to Germany or Italy: *When fascism comes to America it will be wrapped in the flag and carrying a cross* (Sinclair Lewis).

Nor have communism and socialism developed effective resistance to the destruction of the terrestrial environment, in part due to their focus on materialism and living standards and in part ignorance of physical science and thereby the future of life.

In so far as it has been assumed that the appropriately named MAD (Mutual Assured Destruction) policies may avert wars, the re-emergence of fascist parties around the world confounds such hopes. With time a possible nuclear accident becomes a probability and consequently a certainty. Underlying repeated orgies of death, called *war,* represent: (1) Pent-up anger, an atmosphere conducive to violence and revenge against perceived internal or external enemies, commonly incited by ruling elites and their media mouthpieces; (2) An "economic" need for maintaining an ever-growing military-industrial complex and fossil fuel industries; (3) Global propaganda and entertainment industries that disseminate an anti-culture of crass attitudes, obscenities and violence. The eternal battle between life and death, repeating itself every several decades—the 30 year war, the Napoleonic wars, WWI, WWII, Korea, Vietnam, the Middle East—instigated by tyrannical rulers, religious zealots and merchants of death—does not appear to be capable of ending. Advocates of survival, including humanitarians are frozen out from media platforms and in some cases eliminated.

In the words of Joachim Hans Schellnhuber, Europe's chief climate scientist, +4.0 degrees Celsius signifies the breakdown of civilization. At a time of unprecedented global emergency the re-emergence of neo-fascism around the world prevents both, a dismantling of the nuclear genocide machine and attempts at climate mitigation and adaptation. Betrayal is everywhere. There was a time when a Prime Minister stated: *Now our response to climate change must be guided by science. The science tells us that we have already exceeded the safe upper limit for atmospheric carbon dioxide. We are as humans conducting a massive science experiment with this planet. It's the only planet we've got"*, subsequently presiding over a pro-coal government. There was a time a party regarded climate change as "*the greatest moral challenge of our time*", ending up in support of coal mining and export.

The enlightenment, defined as "*ideas centered on reason as the primary source of authority and legitimacy, advancing ideals like liberty, progress, tolerance, fraternity, constitutional government and separation of church and state*", rising above prejudices and witchcraft, is in full retreat. As the late Patrick White mourned, had even a fraction of the tens of thousands of sport carnival spectators taken part in peace rallies, the world could have changed. More recently, whereas the evidence for an abrupt shift in state of the atmosphere has become manifest, a space cult is emerging promising an escape—an interplanetary future for humanity in barren space (Fig. 10.5).

10

Space Lunacy

To ignore evil is to become an accomplice (Martin Luther King).

History testifies to the open-ended ambitions of powerful rulers for deification as gods, including kings, Pharaohs and Emperors such as Caligula (Fig. 10.1) or Nero, mimicked by billionaires and their followers, including some scientists, by false messianic prophecies of "intergalactic civilization" (Fig. 10.2).

With the onset of space exploration, from the Sputnik to the lunar landings and the exploration of Mars, religious mythologies evolve into space cults, alluding to a colonization of planets and a spread of human civilization to space where, presumably, new environments would be overwhelmed by the species. Predictions of making life interplanetary by billionaire proprietors of space hardware enterprises, such as Elon Musk, Jeff Bezos and Richard Branson, including plans for space tourism, asteroid mining (Odenwald 2017) and permanent human settlements on the Moon and Mars (Fig. 10.5) would by some estimates be expected to cost about $1 trillion by 2040. In 2000 Jeff Bezos, Amazon founder and the world's richest man, has launched the reusable Blue Origin, albeit with several launch engine problems. The aim is to commence space tourism in sub-orbital flights, charging a six-figure price such as $300,000 per ticket. Further developments of returnable rockets include defense contracts with the US government and ambitions for permanent human settlement on the Moon, in partnership with NASA.

In 2002 Elon Musk, founder of Pay-Pal, developed the SpaceX rocket, including 70 launches to date, signing contracts with NASA, the US Air

A. Y. Glikson, *The Event Horizon: Homo Prometheus and the Climate Catastrophe*,
https://doi.org/10.1007/978-3-030-54734-9_10

Fig. 10.1 Bust of Caligula Roman emperor from 37 to 41 A.D., from Palazzo Massimo. *Public domain*

Force and the Argentine Space Agency, including supply contract with the International Space Station. Space-X's ultimate goal is to send crewed flights to Mars and eventually colonize the Red Planet. "*I want to die on Mars,*" Musk has said (2017), "*just not on impact.*", with the motto being: "*Making Life Interplanetary*" ... "*You want to wake up in the morning and think the future is going to be great—and that's what being a spacefaring civilization is all about. It's about believing in the future and thinking that the future will be better than the past. And I can't think of anything more exciting than going out there and being among the stars.*"

In 2004 Richard Branson launched *Virgin Galactic,* a tourist-oriented reusable 'space plane' for sub-orbital flights, having already signed some rich people on $250,000 tickets and collaborating with the UAE's sovereign wealth fund. On 13 December 2018 the VSS Unity achieved the project's first suborbital space flight, reaching an altitude of 82.7 km. In February 2019 a member of the team sat in a flight that reached an altitude of 89.9 km. In Richard Branson's Motto "*Together we open space to change the world for good*".

Fig. 10.2 **a** Adam and Eve; **b** Homo sapiens on the Moon

These ideas have been attracting some scientists like bees to the honey-pot. Stephen Hawking (Fig. 10.3) said: "*Human race is doomed if we do not colonize the Moon and Mars*" (Knapton 2017), and further stated: "*I am convinced that humans need to leave earth. The Earth is becoming too small for us, our physical resources are being drained at an alarming rate*" and "*We have given our planet the disastrous gift of climate change, rising temperatures, the reducing of polar ice caps, deforestation and decimation of animal species.*"

It is inconceivable that Albert Einstein, the greatest scientist and humanist, would relegate the human future to space. It appears that, just as the masters of carbon and nuclear energies are willing to sacrifice life, some scientists propose to recreate life on other planets, God-like, regardless of the physics of these planets' atmosphere and the biology of the human body. Arising from Hawking's assertions is the question whether humans will do to the Martian and Lunar environments what they are doing to planet Earth? Space prophets include mainly physicists, but very few biologists, and do not understand that the human body and psychology are inexorably connected with the Earth.

Fig. 10.3 Stephen Hawking with the background of the Horse Nebula, NASA *Public domain*

Scientific exploration of the planets is an exciting idea, but the $trillions to be committed for this purpose come from the mouths of hungry children. In the background of space games is the rise of climate disruption, the upsurge of fascism and potential nuclear calamities, of which proponents of space colonization take little account. Sacrificing the living Earth, all humans may be left with are a few barren space rocks to temporarily support a few survivors.

We are Earthlings, our bodies evolved on Earth, attuned to its gravity, atmosphere, radiation and the multitude of micro-organisms on whom we depend. Prophecies of space colonization imply that alternatives exist to attempts to save the terrestrial atmosphere, oceans and biosphere.

Exploration of the planets best belongs to mobile robotic micro-laboratories designed to monitor the electromagnetic wave spectrum. Nor can space prophets explain how, should humans succeed in colonizing one or more of the planets, their psychology would change in terms of internal conflicts and environmental destruction (Morton, 2018). In H.G. Wells' *War of the Worlds* the Martians have become warlike, invading Earth (Fig. 10.4).

According to Oxfam eight billionaires now own as much wealth as half the human race, playing with space toys out of this planet. In an ethics-free

Fig. 10.4 An army of Martian fighting-machines destroying England (1906) (From H.G. Wells (1895) *War of the Worlds*). *Public Domain*

age false prophecies of planetary colonization (Fig. 10.5)—of the rich, by the rich, for the rich—can only constitute a diversion from the desperate need to save life on Earth. The parallels between religious beliefs of heaven and hell are evident, the virtuous. i.e. the ultra-rich, would be salvaged while the poor and colored skin will burn in hell, as Earth is warming.

It is not possible to argue with insanity.

Fig. 10.5   An artist's concept of a lunar colony

11

Notes from a Catastrophe

Let us not talk falsely now, the hour is getting late (Bob Dylan).

Australia is burning. Following a long drought the country is tinder dry, it only takes a spark, from lightning, power shortcut, a match or a powerful light source to ignite near to 190,000 km^2 of the natural bush and national parks. For a country mining and exporting more coal than any these fires manifest the absurd in the sense of Albert Camus and his hero Sisyphus (Fig. 11.1). By 14 January 2020 fires have burnt an estimated 186,000 square kilometers (Fig. 11.2) and more than 2,779 homes, killing at least 34 people and an estimated one billion animals, with endangered species driven toward extinction. As "authorities" talk about hazard reduction (Kinsella and Jackson 2020), it appears the forests themselves have become a *"Hazard"*, to be reduced even further, culminating the species' thousands of years-long war against the trees and against nature itself (France-Presse 2019).

Untruths are covered up by the media. Countries supposed to switch to clear alternative energy mine are drilling, mining and exporting fossil fuels all over the world, as if the atmosphere can discriminate between domestic and exported greenhouse gases. As the warming oceans keep rising *Homo sapiens* proceeds to alter the composition of the atmosphere, transferring every accessible carbon molecule from the ground to the air, generating fires in an auto-da-fe of the living Earth, pouring $trillions into preparations for a nuclear war. People scuttle like ants before the rain waving "busy" flags, jet around the world to visit *"the top places to see before it is too late"*, frequent

A. Y. Glikson, *The Event Horizon: Homo Prometheus and the Climate Catastrophe*, https://doi.org/10.1007/978-3-030-54734-9_11

Fig. 11.1 Humpty-Dumpty sat on a wall, Humpty-Dumpty had a great fall. From the 1902 Mother Goose story book by William Wallace Denslow. *Public domain*

mass circuses where *sport legends, World-Number-One Icons, "Academy Award"* winners and assorted celebs write themselves into history.

As the rising oceans are saturated with plastic particles that kill marine life and birds, the Amazon burns, remaining Australian forests are burning, Brazil's coast is inundated by huge oil spills, the consequences of the extraction and transfer of carbon into the atmosphere and microplastic particles into the hydrosphere, poisoning the marine food chain, are increasingly manifest. Champions of fossil fuel dance on the graves of nature: Captain Star, a racing ace and aerobatics pilot of right stuff fame, prides himself on burning the liquid residues of plankton that populated Jurassic Tethys Sea some 150 million years ago. Shock jocks and talking heads prescribe untruths, snake oil merchants numb the young, empires cast surveillance nets, arms dealers and fossil fuel tycoons pour $billions into elections.

Unwittingly the computers have taken control, of communication systems, of the share markets, of on-alert missile launch systems that can be activated

Fig. 11.2 **a** Satellite image of fires along the southern coast of New South Wales, NASA; **b** The Orroral Valley Fire viewed from Tuggeranong in southern Canberra. *Permission*

by a faulty computer chip. The Morons-in-charge dominate the airwaves. A fat senator declares: "*To be quite honest, you know, it's the twenty-first century, you know, it's a challenge, you know, we just need to grow the economy going forward, you know*".

Shrieking sounds and primordial howls that pass for music betray an unconscious knowledge by the youth of what's to come. Half naked long haired boys and girls sing about life-long love. Young people wearing earphones, thumbs on smart phones, stumble into video shops. There images of black knights protecting blond girls aim guns and knives in a background of stars and stripes. Two leather-clad chain- nearing swastika-adorned bike riders pass by—tomorrow's killers waiting for the order. Memories of Weimar Brown Shirts re-emerge.

Fig. 11.3 Clock of the Atomic Scientists: It is 100 s to mid-night. *Public domain*

Can the consequences of the great carbon oxidation event and a nuclear winter combine in a mass extinction on a Permian–Triassic scale some 250 million years ago, when some 96 percent of marine species and 70 percent of terrestrial species were lost due to oxygen depletion and release of toxic hydrogen sulphide from the deep ocean (Kump et al. 2005; Penn et al. 2018; Hickey 2018)?

It is hard to think of a science fiction describing a civilization which, against the best scientific evidence, proceeds knowingly to destroy its own life support systems. People can only worry about one issue at a time. A nuclear conflict may occur by accident or design (Elsberg 2017) at any moment, even before the worst consequences of global warming are realized. Humans may survive, but will they be able to change in time (Fig. 11.3)?

12

A Eulogy

Laugh at us all … you and the others who sit with you, grinning fools at the entrance to hell. Laugh and laugh as the ash falls soft as silent despair (Richard Flanagan).

A voice has emerged on behalf of future generations through Greta Thunberg (Fig. 1.1) who, like the child in Hans Christian Andersen's story, declared "*The Emperor has no cloth*".

Overwhelmed by the beauty of the world permeating all nature, all the more daunting because of its transience, as it has always been, and the unthinkable crime that "*sapiens*" is committing,

As for the monster – well, it is the transitory nature of things that makes them beautiful. The only things that don't change are dead things. The silver lining is that the world seems so much more beautiful when we know that we are saying goodbye to it, at least to the way we know it (Judith Crispin).

When I thought '*sometimes I imagine I am dreaming the world*', Tom responded saying '*sometimes I wonder whether the world is not dreaming us*'.

© The Editor(s) (if applicable) and The Author(s), under exclusive
license to Springer Nature Switzerland AG 2021
A. Y. Glikson, *The Event Horizon: Homo Prometheus and the Climate Catastrophe*,
https://doi.org/10.1007/978-3-030-54734-9_12

Dreaming a UniVerse

Had eternity been true
Had time never began
Why in this life do I meet you
And not in any other span?
Had the universe been boundless
Were we lost in legion space
How come you live in this address
But not in any other place?
Had there been a first creator
Of the grand cosmic design
Who is the well concealed narrator?
Who spied on the divine?
Then some proclaim they own
Truth, by secret revelation
In ignorance of the unknown
Closed minds conceive creation.
But as to me, I marvel birth
Of tiny buds from kernel seeds
How sprouting life blooming on Earth
Despite of entropy succeeds.
How did the primeval cell know
It will grow to become you?
What forms will you take in flow
When one billion years are due?
I will meet you there!

Magpies, Currawongs, Kookaburras and possums around me are fortunate. They do not know and therefore cannot worry about the future, while me, Cassandra-like (Fig. 12.1) am overcome by grief. Humans are an offspring of nature, yet are destroying nature, but what we are perpetrating can only be inherent in nature. Mass extinctions of species, mainly due to external factors such as volcanism and asteroid impacts, have occurred in the past (Stanley 1987; Saltre and Bradshaw 2019) but ours is a different story. Humans can be beautiful but not entirely sane, manipulated by life-negating forces that commit wars, genocide and mass extinction. Staring at the unthinkable we live at a moment of truth for us and the species we are bringing down with us.

Having lost the sense of reverence toward the Earth possessed by prehistoric humans, there is no evidence civilization is about to adopt Carl Sagan's

Fig. 12.1 Cassandra by Evelyn De Morgan (1898, London); Cassandra in front of the burning city of Troy. *Public domain*

(1980) sentiment: '*For we are the local embodiment of a cosmos grown to self-awareness. We have begun to contemplate our origins: star stuff pondering the stars: organized assemblages of ten billion billion billion atoms considering the evolution of atoms; tracing the long journey by which, here at least, consciousness arose. Our loyalties are to the species and the planet. We speak for the Earth. Our obligation to survive is owed not just to ourselves but also to that Cosmos, ancient and vast, from which we spring.*'

Humans live in a realm of perceptions, dreams, myths and legends, in denial of realities and critical facts (Koestler, 1978). They wake up for a brief moment from an infinite universal slumber to witness a world as cruel as it is beautiful, a biosphere dominated by the food chain. An inverse relation may exist between the level of consciousness achieved by a species and its longevity, once it creates machines and processes it cannot control. If looking

Fig. 12.2 Earthrise 1 Historic Image. Credit: NASA, Apollo 8 Crew. *Permission*

into the sun may result in blindness so, according to little-understood laws, the deep insights into nature humans have achieved may bear a terrible price.

Existentialist philosophy allows a perspective into, and a way of coping with, what defies rational contemplation. Ethical and cultural assumptions of free will rarely govern the behavior of societies, let alone an entire species. And although the planet may not shed a tear for the demise of technological civilization, hope, on the individual scale, is still possible in an existential sense. Going through their black night of the soul, members of the species may be rewarded by the emergence of a conscious dignity devoid of illusions, grateful for the glimpse at the universe for which humans are privileged by the fleeting moment:

"*Having pushed a boulder up the mountain all day, turning toward the setting sun, we must consider Sisyphus happy.*" (Camus 1942) (Fig. 12.2).

Because We Believe

Once in every life there comes a time
We walk out all alone and into the light
The moment won't last, but then
We remember it again,
When we close our eyes
Like stars across the sky
We were born to shine
Because we believe

Andrea Bocelli

References

Adler J (2013) Fire made us human. Smithsonian Magazine, June 2013

Alvarez LW et al (1980) Extra-terrestrial cause for the cretaceous-tertiary extinction: experimental results and theoretical interpretation. Science 208:1085–1095

Andrews RG (2019) Can spaceflight save the planet? Sci Am, September 6, 2019

Baek-Min K et al (2014) Weakening of the stratospheric polar vortex by Arctic sea-ice loss. Nature Comm 5, Article 4646

Barlow J, Lees AC (2019) Amazon fires explained: what are they, why are they so damaging, and how can we stop them? The Conversation, August 24, 2019

Barnosky AD et al (2012) Approaching a state shift in Earth's biosphere. Nature 486:52–58

Beerling DJ et al (2002) An atmospheric pCO_2 reconstruction across the Cretaceous-Tertiary boundary from leaf mega-fossils. Proc Natl Acad Sci USA 99(12):7836–7840

Beerling DJ, Royer D (2011) Convergent cenozoic CO_2 history. Nat Geosci 4:418–420

Bengtsson L (2006) Storm tracks and climate change. J Clim 19:15

Berger A, Loutre MF (2002) An exceptionally long interglacial ahead? Science 297:1287–1288

Berkeley Earth (2018) 4th Hottest year on record. http://berkeleyearth.org/summary-of-findings/

Berner RA (2004) The phanerozoic carbon cycle: CO_2 and O_2. Oxford University Press, Oxford, p 160

Berner RA (2006) GEOCARBSULF: a combined model for Phanerozoic atmospheric O2 and CO2. Geochim et Cosmochim Acta 70:5653–5664

© The Editor(s) (if applicable) and The Author(s), under exclusive license to Springer Nature Switzerland AG 2021
A. Y. Glikson, *The Event Horizon: Homo Prometheus and the Climate Catastrophe*, https://doi.org/10.1007/978-3-030-54734-9

Betzler C et al (2018) Refinement of Miocene sea level and monsoon events from the sedimentary archive of the Maldives (Indian Ocean). Prog Earth Planet Sci 5:5

Biro D et al (2013) Tool use as adaptation. Philos Trans R Soc Lond B Biol Sci. 368(1630):20120408

Bowman DM (2009) Fire in the Earth system. Science 24, 324(5926):481–484

Braun H et al (2005) Possible solar origin of the 1470-year glacial climate cycle demonstrated in a coupled model. Nature 438:208–211

Broecker WS (2000) Abrupt climate change: causal constraints provided by the paleoclimate record. Earth Sci Rev 51:137–154

Broecker WS, Stocker TF (2011) The holocene CO_2 rise: anthropogenic or natural? Eos 87(3)

Bronselaer B et al (2018) Change in future climate due to antarctic meltwater. Nature 564:53

Butler JH, Montzka SA (2019) The NOAA annual greenhouse gas index (AGGI). MOAA Earth System Research Laboratory Global Monitoring Division. https://www.esrl.noaa.gov/gmd/aggi/aggi.html

Camus A (1942) The myth of sisyphus. Penguin Classics 2000

Canadell P (2009) Super-size deposits of frozen carbon threat to climate change. Global Biogeochemical Cycles. Eureka Alert. https://www.globalcarbonproject.org/

Carrington D (2019) Climate crisis: 11000 scientists warn of 'untold suffering'. The Guardian, Wed 6 November

Cartmill M, Smith FH (2009) The human lineage. Wiley-Blackwell, Hoboken

Cartwright M (2013) Prometheus. Ancient history encyclopedia https://www.ancient.eu/Prometheus/

Cartwright M (2018) Aztec sacrifice. Ancient history encyclopedia https://www.ancient.eu/Aztec_Sacrifice/

Ceballos G et al (2015) Accelerated modern human—induced species losses: entering the sixth mass extinction. Sci Adv 1(5):e1400253

Chandler M et al (2008) The PRISM model/data cooperative: mid-pliocene data-model comparisons. Pages News 16(2):24–25

Charbit S et al (2008) Amount of CO_2 emissions irreversibly leading to the total melting of Greenland. Geophys Res Lett 35:12

Conway EM, Oreskes N (2010) Merchants of doubt. Angus and Robertson

Cooperman Y (2016) Biochar and Carbon Sequestration. Solution centre for nutrient management: Perspectives on nutrient management in California agriculture

Cortese G et al (2007) The last five glacial-interglacial transitions: a high resolution 450,000-year record from the sub-antarctic atlantic. Paleo-Ocean 22:PA4203

Covey C et al (1994) Global climatic effects of atmospheric dust from an asteroid or comet impact on Earth. Glob Planet Change 9:263–273

Cui Y et al (2011) Slow release of fossil carbon during the Paleocene-Eocene Thermal Maximum. Nat Geosci 4:481–485

Dakos V et al (2008) Slowing down as an early warning signal for abrupt climate change. Proc Nat Assoc Sci 105(38):14308–14312

De Menocal PB (2004) African climate change and faunal evolution during the Pliocene-Pleistocene. Earth Planet Sci Lett 220:3–24

DeConto RM, Pollard D (2003) Rapid Cenozoic glaciation of Antarctica induced by declining atmospheric CO_2. Nature 421.245–249

Denton GH (2010) The Last Glacial Termination. Science 328:1652–1656

Dergachev VA, Raspopov OM (2000) The solar cycle and variation of the global air temperatures since 1868. ESA Publications Division, 2000 xi, 680 p. Eu Space Agen, Vol. 463, p.485

Drum K (2020) We need a massive climate war effort—now: only major spending on clean energy R&D can save us. Mother Jones. Jan;Feb 2020 issue

Elsberg D (2017) The doomsday machine: confessions of a nuclear war planner. Bloomsbury Press, 387 pp

Eby M et al (2009) Lifetime of anthropogenic climate change: millennial time scales of potential CO_2 and surface temperature perturbations. J Clim 22:2501–2511

Environmental Migration Portal (2015) Regional Maps on Migration, Environment and Climate Change. https://environmentalmigration.iom.int/maps

EPICA community members (2004) Eight glacial cycles from an Antarctic ice core. Nature 429:623–628

Flanagin J (2015) Aztec human sacrifice was a bloody, fascinating mess. https://qz.com/374994/aztec-sacrifice-was-real-and-its-not-fetishistic-to-be-fascinated-by-it/

Foster GL, Rohling EJ (2013) Relationship between sea level and climate forcing by CO_2 on geological timescales. Proc Nat Assoc Sci 110(4):1209–1210

France-Presse Agence (2019) UN Chief: Humanity's 'War against Nature' Must Stop. https://www.voanews.com/europe/un-chief-humanitys-war-against-nature-must-stop

Freeman D (2019) NASA video shows polar vortex shifting to spread frigid air across the US. Mach Feb. 2, 2019. https://www.nbcnews.com/mach/science/nasa-video-shows-polar-vortex-shifting-spread-frigid-air-across-ncna965951

Ganopolski A, Rahmstorf S (2001) Rapid changes of glacial climate simulated in a coupled climate model. Nature 409(6817):153–158

Ganopolski A et al (2016) Critical insolation-CO_2 relation for diagnosing past and future glacial inception. Nature 534:S19–S20

Gehler AQ et al (2016) Temperature and atmospheric CO2 concentration estimates through the PETM using triple oxygen isotope analysis of mammalian bioapatite. Proc Natl Acad Sci USA 113(28):7739–7744

Glikson AY (2013) Fire and human evolution: the deep-time blueprints of the Anthropocene. Anthropocene 3:89–92

Glikson AY (2016) Cenozoic greenhouse gases and temperature changes with reference to the Anthropocene. Glob Change Biol 22(12):3843–3858

Glikson AY, Groves CP (2016) Climate, fire and human evolution: the deep time dimensions of the anthropocene. Springer, 176 p

Glikson AY (2019) North Atlantic and sub-Antarctic Ocean temperatures: possible onset of a transient stadial cooling stage. Climat Change, Springer, 155/3:311–321

Global Carbon Budget (2018). Global carbon project. https://doi.org/10.18160/gcp-2018

Golledge NR (2019) Global environmental consequences of twenty-first-century ice-sheet melt. Nature 566(7742):65–67

Gopnik A (2012) Facing history: why we love Camus. The New Yorker. April 9, 2012 issue

Groves CP (2016) cenozoic biological evolution. In: Glikson AY, Groves CP (eds) Climate, fire and human evolution the deep time dimensions of the Anthropocene. Springer, 226 pp

Guminer Y (1931) Arithmetic of a counter-plan poster https://commons.wikimedia.org/wiki/File:Yakov_Guminer_-_Arithmetic_of_a_counter-plan_poster_(1931).jpg

Hansen JE et al (2007) Climate change and trace gases. Phil Trans Roy Soc 365A:1925–1954

Hansen JE (2007b) Scientific reticence and sea level rise. Environ Res Lett 2/024002 (6 pp)

Hansen JE et al (2008) Target atmospheric CO_2: where should humanity aim? Open Atmos Sci J. 2:217–231

Hansen JE et al (2011) Earth's Energy Imbalance and Implications. Atmos Chem Phys 11:13421–13449

Hansen JM, Sato M (2012) Paleoclimate implications for human-made climate change. Clim Change 21–48

Hansen JE et al (2013a) Climate sensitivity, sea level, and atmospheric carbon dioxide. Phil Trans Roy Soc A371, Issue 2001

Hansen JM et al (2013b) doubling down on our faustian bargain. http://www.columbia.edu/~jeh1/mailings/2013/20130329_FaustianBargain.pdf

Hansen JE et al (2013c) Assessing "Dangerous Climate Change": required reduction of carbon emissions to protect young people, future generations and nature. PLoS ONE 8(12):e81648

Hansen JE et al (2016) Ice melt, sea level rise and superstorms: evidence from paleoclimate data, climate modeling, and modern observations that 2 °C global warming could be dangerous. Atmos Chem Phys 16:3761–3812

Hansen JE (2018) Climate change in a nutshell: the gathering storm. http://www.columbia.edu/~jeh1/mailings/2018/20181206_Nutshell.pdf

Hickey H (2018) What caused Earth's biggest mass extinction? Stanford University School of Earth, Energy and Environmen Science. https://earth.stanford.edu/

Hoare C (2018) NASA asteroid warning - why space rock could spark 'cosmic winter' on Earth https://www.express.co.uk/news/science/1118874/nasa-asteroid-warning-space-rock-cosmic-winter-earth-spt

Hudson SR (2011) Estimating the global radiative impact of the sea ice–albedo feedback in the Arctic. J Geophys Res 116:1–7

Kim Intae et al (2016) The distribution of glacial meltwater in the Amundsen Sea, Antarctica, revealed by dissolved helium and neon. J Geophys Res Oceans 121:1654–1666

IPCC-AR4 (2007) Contribution of working group I to the 4th assessment report. The Intergovernmental Panel of Climate Change. https://www.ipcc.ch/assess ment-report/ar4/

IPCC-AR5 (2013) Managing the risks of extreme events and disasters to advance climate change adaptation. Special report of the intergovernmental panel on climate change https://www.ipcc.ch/site/assets/uploads/2018/03/SREX_Full_ Report-1.pdf

IPCC-AR5 (2018) Drivers, trends and mitigation. ipcc.ch/site/assets/uploads/2018/ 02/ipcc_wg3_ar5_chapter5.pdf

Jones N (2019) Polar warning: even Antarctica's coldest region Is starting to melt. Yales Environ 360

Jouzel J et al (2007) Orbital and millennial Antarctic climate variability over the past 800,000 years. Science 317:793–796

Keeling RF et al (2008) Atmospheric carbon dioxide record from Mauna Loa. CDIAC. Carbon Dioxide Research Group, Scripps Institution of Oceanography University of California

Keller G (2005) Impacts, volcanism and mass extinction: random coincidence or cause and effect? Austr J Earth Sci 52(4):725–757

Kennett JP, Stott LD (1991) Abrupt deep-sea warming, palaeoceanographic changes and benthic extinctions at the end of the Palaeocene. Nature 353:225–229

Kidston J (2012) Poleward-shifting climate zones—where are they headed and why? The Conversation, May 28

Kinsella E, Jackson W (2020) What are hazard reduction burns, are we doing enough of them, and could they have stopped Australia's catastrophic bushfires? ABC news 10 Jan 2020

Knapton S (2017) Human race is doomed if we do not colonize the Moon and Mars, says Stephen Hawking. The Telegraph. https://www.telegraph.co.uk/ science/2017/06/20/human-race-doomed-do-not-colonise-moon-mars-says-ste phen-hawking/

Koestler A (1978) Janus: A summing up. https://www.abebooks.com/first-edition/ JANUS-Summing-Arthur-Koestler-Hutchinson/13362028523/bd. Hutchinson, 354 pp

Komninos A (1998) Our three brains—the reptilian brain. Interaction Desiign Foundation.

Kring DA (2000) Impact events and their effect on the origin, evolution, and distribution of life. GSA Today 10/8

Kring D (2002) Chicxulub impact event. Lunar and Planetary Institute. https:// www.lpi.usra.edu/science/kring/Chicxulub/regional-effects/

Klein RG, Edgar B (2002) The dawn of human culture. Wiley—Social Science, 288 pp

Kump LR et al (2005) Massive release of hydrogen sulfide to the surface ocean and atmosphere during intervals of oceanic anoxia. Geology 33(5):397–400

Kump LR (2011) The Last Great Global Warming. Sci Am. July 2011

Kurschner WM et al (2008) The impact of Miocene atmospheric carbon dioxide fluctuations on climate and the evolution of terrestrial ecosystems. Proc Nat Acad Sci 15:449–453

Kutzbach JE et al (2010) Climate model simulation of anthropogenic influence on greenhouse-induced climate change (early agriculture to modern): the role of ocean feedbacks. Clim Change 99:351–381

Lashof D (2018) Why positive climate feedbacks are so bad. World Resources Institute. https://www.wri.org/blog/2018/08/why-positive-climate-feedbacks-are-so-bad

Leber R (2015) Obama is right: climate change kills more people than terrorism. The New Republic. February 12, 2015

Lenton TM et al (2008) Tipping elements in the Earth's climate system. Proc Nat Acad Sci 105(6):1786–1793

Levermann A, Mengel M (2014) Tipping point for Antarctic melting. Nature 509:136

Lindsay R (2019) Climate change: atmospheric carbon dioxide. NASA Climate Gov. https://www.climate.gov/news-features/understanding-climate/climate-change-atmospheric-carbon-dioxide

Liu Z et al (2009) Global cooling during the Eocene-Oligocene climate transition. Science 323:1187–1190

Londono L et al (2018) Early Miocene CO_2 estimates from a Neotropical fossil leaf assemblage exceed 400 ppm. Am J Botany 105(11):1929–1937

Lucas C et al (2014) The expanding tropics: a critical assessment of the observational and modeling studies. WIRES climate change. Roy Geogr Soc 5(1):89–112

Lucht W (2002) Climatic control of high-latitude vegetation greening trend and Pinatubo effect. Science 296(5573):1687–1689

Malouf D (1996) Epimetheus, or the spirit of reflection. Australian Humanities Review 65

Mann ME et al (1999) Northern hemisphere temperatures during the past millennium: Inferences, uncertainties, and limitations. Geophys Res Lett 26:759–762

Maslin MA, Brierley CM, Milner AM, Shultz S, Trauth MH, Wilson KE (2014) East African climate pulses and early human evolution. Quat Sci Rev 101:1–17

Matsuzawa T (2012) What is uniquely human? A view from comparative cognitive development in humans and chimpanzees. In: De Waal FBM, Ferrari PF (eds) The primate mind. Harvard University Press, Cambridge, pp 288–305

McInerney FA, Wing SL (2011) The paleocene-eocene thermal maximum: a perturbation of carbon cycle, climate, and biosphere with implications for the future. Ann Rev Earth Planet Sci 39:489–516

McSweeney R (2015) Warming oceans less able to store organic carbon, study suggests. Carbon Brief, 6 January 2015.

Miller KG et al (2012) High tide of the warm Pliocene: Implications of global sea level for Antarctic deglaciation. Geology 40/5

Morton A (2018) Colonizing other planets could trigger war on Earth. Newsweek 11/22/18.

Morton J (2018) Don't mention the emergency: making the case for emergency climate action. https://climateemergencydeclaration.org/wp-content/uploads/2018/09/DontMentionTheEmergency2018.pdf

Mouginot J et al (2019) Forty-six years of Greenland Ice Sheet mass balance from 1972 to 2018. Proc Nat Acad Sci 116(19):9239–9244

Mumford L (1972) Lewis Mumford, The Human Heritage. https://sniadecki.wordpress.com/2012/04/11/mumford-heritage-en/

Musk E (2017) Making Life interplanetary. https://www.spacex.com/sites/spacex/files/making_life_multiplanetary_transcript_2017.pdf

NASA (2014) World of change: global temperatures. https://earthobservatory.nasa.gov/world-of-change/DecadalTemp

NASA (2019) A new global fire atlas. October 22 https://earthobservatory.nasa.gov/images/145417/a-new-global-fire-atlas

National Academy Press (2011) Understanding earth's deep past: lessons for our climate future. Chapter 3: climate transitions, tipping points, and the point of no return. https://www.nap.edu/read/13111/chapter/6

National Centre Atmospheric Research (NCAR) (2011). Melting arctic sea ice and ocean circulation. https://scied.ucar.edu/longcontent/melting-arctic-sea-ice-and-ocean-circulation

Naughton J (2018) Anti-Social Media: How Facebook Disconnects Us and Undermines Democracy by Siva Vaidhyanathan—review. The Guradian, Mon 25 Jun 2018

New M et al (2011) Four degrees and beyond: the potential for a global temperature increase of four degrees and its implications. Phil Trans Roy Soc. 369, Issue 1934.

NOAA (2019) Annual mean growth rate for Mauna Loa, Hawaii. NOAA Earth System Research Laboratory Global Monitoring Division. https://www.esrl.noaa.gov/gmd/ccgg/trends/gr.html

NOAA (2019) Global warming and hurricanes: an overview of current research results. https://www.gfdl.noaa.gov/global-warming-and-hurricanes/

Notro M et al (2006) Global vegetation and climate change due to future increases in CO_2 as projected by a fully coupled model with dynamic vegetation. Center for Climatic Research, University of Wisconsin—Madison, Madison, Wisconsin

Odenwald S (2017) The Myth of Space Mining. Huffington Post. Dec 06, 2017

Oreskes N, Michael Oppenheimer (2019) Scientists have been underestimating the pace of climate change. Sci Am 175

Overpeck JT et al (2006) Paleoclimatic evidence for future ice-sheet instability and rapid sea-level rise. Science 311(5768):1747–1750

Orwell G (1950/2011) 1984. Penguin Books. 354 pp

Quintero I, Wiens JJ 2013. Rates of projected climate change dramatically exceed past rates of climatic niche evolution among vertebrate species. Ecol Lett 16/8

Pagani M et al (2006) Arctic hydrology during global warming at the Paleocene/ Eocene thermal maximum. Nature 442(10):671–675

Pekar S, Christie-Blick N (2007) Showing a strong link between climatic and pCO_2 changes: resolving discrepancies between oceanographic and Antarctic climate records for the Oligocene and early Miocene (34–16 Ma) Geological Survey and the National Academies; USGS OF-2007-1047, Extended Abstract 024.

Penn JL et al (2018) Temperature-dependent hypoxia explains biogeography and severity of end-Permian marine mass extinction. Science 362(6419)

Petit JR et al (1999) 420,000 years of climate and atmospheric history revealed by the Vostok deep Antarctic ice core. Nature 399:429–436

Pickrel J (2018) Australian raptors start fires to flush out prey. Cosmos News Biology, 12 January 2018

PIK (2012) Four-degrees briefing for the World Bank: The risks of a future without climate policy. Potsdam Institute for Climate Impact Research (PIK) https://www.pik-potsdam.de/news/press-releases/archive/2012/4-degrees-briefing-for-the-world-bank-the-risks-of-a-future-without-climate-policy

Piper K (2019) Death by algorithm: the age of killer robots is closer than you think. Vox, 21 June, 2019

Pyne SJ (2016) Fire in the mind: changing understandings of fire in Western civilization. Philos Trans R Soc Lond B Biol Sci. 2016 Jun 5; 371(1696)

Rahmstorf SR, Coumou D (2011) Increase of extreme events in a warming world. Proc Natl Acad Sci USA 108:17905–17909

Rahmstorf S et al (2015) Exceptional twentieth-century slowdown in Atlantic Ocean overturning circulation. Nature Climate Change 5:475–480

Rahmstorf S, Stocker TF (2004) Thermohaline circulation: past changes and future surprises? The IGBP series 2005, pp 240–241. http://www.pik-potsdam.de/~Stefan/Publications/Book_chapters/rahmstorf&stocker_2004.pdf

Renne PR et al (2013) Time scales of critical events around the Cretaceous-Paleogene boundary. Science 339(6120):684–687

Rignot E et al (2019) Four decades of Antarctic Ice Sheet mass balance from 1979–2017. Proc Natl Acad Sci USA 116(4):1095–1103

Ritchie H, Roser (2019) CO_2 and greenhouse gas emissions. World Data, University of Oxford. https://ourworldindata.org/co2-and-other-greenhouse-gas-emissions

Roberts N (1998) The Holocene. Blackwell Publishers, Oxford. https://www.amazon.com/Holocene-Environmental-History-Neil-Roberts/dp/0631186387

Royer DL et al (2001) Phanerozoic atmospheric CO change: evaluating geochemical and paleo-biological approaches. Earth-Science Reviews 54:349–392

Royer DL (2006) CO_2-forced climate thresholds during the Phanerozoic. Geochim et Cosmochim Acta 70:5665–5675

Ruddiman WF (1997) Tectonic uplift and climate change. Plenum Press, New York, p 535

Ruddiman WF (2003) Orbital insolation, ice volume, and greenhouse gases. Quat Sci Rev 22:1597–1629

Saltre F, Bradshaw CJA (2019) Are we really in a 6th mass extinction? Here's The Science. The Conversation 18 Nov. 2019.

Sagan C (2002) Cosmos. Random House, 384 pages

Schellnhuber HJ (2009) Tipping elements in the Earth System. Proc Natl Acad Sci USA 106(49):20561–20563

Schoene B et al (2015) U Pb geochronology of the Deccan Traps and relation to the end-Cretaceous mass extinction. Science 347(6218):182–184

Shakhova N et al (2019) Understanding the permafrost-hydrate system and associated methane releases in the east siberian arctic shelf. Geosciences 9:6. https://doi.org/10.3390/geosciences9060251

Shakun JD et al (2012) Global warming preceded by increasing carbon dioxide concentrations during the last deglaciation. Nature (484/7392):49–54

Smith SJ et al (2011) Anthropogenic sulfur dioxide emissions: 1850–2005. Atmos Chem Phys 11:1101–1116

Solomon S et al (2009) Irreversible climate change due to carbon dioxide emissions. Proc Natl Acad Sci USA 106(6):1704–1709

Spinney L (2012) Human cycles: history as science. Nature 488:24–26

Stapledon WO (1972) Last and first man. Penguin, London

Steffen W et al (2018) Trajectories of the Earth System in the Anthropocene. Proc Natl Acad Sci USA 115(33):8252–8259

Steffensen JP et al (2008) High-resolution Greenland ice core data show abrupt climate change happens in few years. Science 32:680–684

Stringer C, Andrews P (2011) The complete world of human evolution (revised edition) Thames & Hudson, London

Su DF, Harrison T (2015) The palaeoecology of the Upper Laetolil Beds, Laetoli Tanzania: a review and synthesis. J African Earth Sci 101:405–419

Thomas C et al (2004) Extinction risk from climate change. Nature 427 (6970):145–148

Turton S (2017) The world's tropical zone is expanding, and Australia should be worried. The Conversation, June 28, 2017

Toon OB et al (1997) Environmental perturbations caused by the impacts of asteroids and comets. Rev Geophys 35:41–78

Yokoyama Y, Esat TM (2015) Global climate and sea level: enduring variability and rapid fluctuations over the past 150,000 years. Oceanography 24(2):54–69

Wagner F et al (2002) Rapid atmospheric CO_2 changes associated with the 8,200-years-B.P. cooling event. Proc Nat Acad Sci USA 99:12011–12014

Wallace-Wells D (2017) The Uninhabitable Earth: Famine, economic collapse, a sun that cooks us: What climate change could wreak—sooner than you think. New York Intelligencer. https://nymag.com/intelligencer/2017/07/climate-change-earth-too-hot-for-humans.html

Watson SK et al (2015) Vocal learning in the functionally referential food grunts of chimpanzees. Curr Biol 25:1–5

Trenberth KE, Shea DJ (2006) Atlantic hurricanes and natural variability in 2005. Geophys Res Lett 33(12)

Welna D (2020) The end may be nearer: doomsday clock moves within 100 seconds of midnight, January 23, 2020

Wells HG (1898) The war of the worlds. William Heinemann Publisher (UK), Pearson's Magazine

Wiersma AP et al (2011) Fingerprinting the 8.2 ka event climate response in a coupled climate model. J Q Sci 26:118–125

Wignall PB, Twitchett RJ (2002) Extent, duration and nature of the Permian—Triassic super-anoxic event. Geol Soc Am Spec Pap 356:395–413

Wilson S (2020) The era of antisocial social media. Harvard Business Review. February 05, 2020

Wolbach WS et al (1990) Major wildfires at the Cretaceous/Tertiary boundary. Geol Soc, Am spec Pap, p 247

Wrangham R (2009) Catching fi re: how cooking made us human. Basic Books, New York, p 320

Wright JD, Schaller MF (2013) Evidence for a rapid release of carbon at the Paleocene-Eocene thermal maximum. Proc Nat Acad Sci 110(40):15908–15913

Zachos et al (2001) Trends, Rhythms, and Aberrations in Global Climate 65 Ma to Present. Science 27(292):686–693

Zachos JC et al (2008) An early Cenozoic perspective on greenhouse warming and carbon-cycle dynamics. Nature 451:279–283

Zeebe RE et al (2009) Carbon dioxide forcing alone insufficient to explain Paleocene-Eocene Thermal Maximum warming. Nature Geoscience 2:576–580

Zeebe RE et al (2016) Anthropogenic carbon release rate unprecedented during the past 66 million years. Nat Geosci 9:325–329

Index

Printed in the United States
by Baker & Taylor Publisher Services